Thermoforming

HOW TO ORDER THIS BOOK

BY PHONE: 800-233-9936 or 717-291-5609, 8AM–5PM Eastern Time

BY FAX: 717-295-4538

BY MAIL: Order Department
Technomic Publishing Company, Inc.
851 New Holland Avenue, Box 3535
Lancaster, PA 17604, U.S.A.

BY CREDIT CARD: American Express, VISA, MasterCard

BY WWW SITE: http://www.techpub.com

PERMISSION TO PHOTOCOPY–POLICY STATEMENT

Authorization to photocopy items for internal or personal use, or the internal or personal use of specific clients, is granted by Technomic Publishing Co., Inc. provided that the base fee of US $3.00 per copy, plus US $ .25 per page is paid directly to Copyright Clearance Center, 222 Rosewood Drive, Danvers, MA 01923, USA. For those organizations that have been granted a photocopy license by CCC, a separate system of payment has been arranged. The fee code for users of the Transactional Reporting Service is 1-56676/98 $5.00 + $ .25.

SECOND EDITION

Thermoforming
A Plastics Processing Guide

DR. G. GRUENWALD, P.E.

TECHNOMIC PUBLISHING CO., INC.
LANCASTER · BASEL

Thermoforming
a **TECHNOMIC**®publication

Published in the Western Hemisphere by
Technomic Publishing Company, Inc.
851 New Holland Avenue, Box 3535
Lancaster, Pennsylvania 17604 U.S.A.

Distributed in the Rest of the World by
Technomic Publishing AG
Missionsstrasse 44
CH-4055 Basel, Switzerland

Copyright © 1998 by Technomic Publishing Company, Inc.
All rights reserved

No part of this publication may be reproduced, stored in a retrieval system, or transmitted, in any form or by any means, electronic, mechanical, photocopying, recording, or otherwise, without the prior written permission of the publisher.

Printed in the United States of America
10 9 8 7 6 5 4 3 2 1

Main entry under title:
    Thermoforming: A Plastics Processing Guide, Second Edition

A Technomic Publishing Company book
Bibliography: p.
Includes index p. 231

Library of Congress Catalog Card No. 98-60402
ISBN No. 1-56676-625-7

Table of Contents

Foreword   ix
Preface   xiii

1. **Introduction** 1

2. **Heating of the Plastic** 3
    Means of conveying heat to the plastic   3
    Physics of radiation heating   5
    Thermal properties of plastics   15
    Heating equipment for plastic sheets   18
    Judging the correct temperature of the heated sheet   27
    Heater controls   31
    Clamping of sheets   33

3. **Thermoforming Molds** 35
    Reduction in wall thickness: male and female molds   35
    Computer-aided engineering for thermoforming   45
    Part shrinkage and dimensional tolerances   48
    Warpage   50
    Draft in the mold   52
    Surface appearance   53
    Mold materials   54
    Mold-cooling provisions   56
    Air passage holes   56
    Increasing stiffness   59
    Mold plugs   59

4. **Vacuum, Air Pressure, and Mechanical Forces** 61
    Measuring vacuum and pressure forces   61
    Vacuum sources   62
    Vacuum accumulators or surge tanks   62
    Application of vacuum forces   64
    Pressure forming   65
    Mechanical forming   66

5. Cooling of Thermoformed Parts 69
Means of cooling the formed part   69
Non-conventional cooling methods   72

6. Trimming of Thermoformed Parts 73
Tools for trimming   74

7. Thermoforming Equipment 79
Single-station thermoformer   80
Shuttle thermoformer   82
Rotary thermoforming equipment   82
Continuous in-line thermoformers   85
In-line thermoformer   97
Linear thermoformers   100
Pneumatic thermoformers   102
Hydraulically operated thermoformers   103
Mechanically operated thermoformers   103
Skin packaging equipment   104
Blister packaging equipment   104
Snap packaging   104
Vacuum packaging   104
Packaging machinery   105
Control mechanisms   109

8. Thermoforming-Related Material Properties 111
Glass transition temperature   111
Heat deflection temperature   112
Softening range and hot strength   112
Specific heat   115
Thermal conductivity   116
Thermal expansion   116
Heat of fusion   117
Thermal diffusivity   118
Thermal stability   118
Water absorption   119
Orientation and crystallization   120
Manufacture of starting materials   128
Coextrusions and laminates   129
Mechanical properties   130
Material economics   131
Regrind utilization   133

9. Thermoforming Materials' Chemical Descriptions 135
Acrylics   136
Cellulosics   136
Polyolefins   137
Styrene polymers   139
Vinyl resins   140
Engineering plastics   141
Copolymers, blends, and alloys   143
Fiber-reinforced thermoplastics   144
Transparent materials   145
Barrier materials   146
Electrical properties   152
Plastics recycling   153
Flammability of plastics   155
Toxicity of plastics   157

10. Thermoforming Processes 159
Billow, bubble, or free forming   159
Cavity forming   161
Drape forming   162
Plug-assist forming   164
Billow drape forming   165
Snap-back forming   166
Air slip forming   167
Reverse draw with plug-assist forming   167
Trapped sheet pressure forming   168
Twin-sheet forming   168
Pressure forming   171
Mechanical thermoforming   172
  Plug-and-ring forming or ridge forming   173
  Slip forming   174
  Matched-mold forming   175
  Rubber pad and fluid pressure or diaphragm forming   175
Other thermoforming processes   175
Adjusting process parameters   177
Thermoforming troubleshooting guide   181

11. Design Considerations 183
Assembly and bonding   185
  Snap-fits   185
  Mechanical bonding   186
  Forming around inserts   186

       Welding   186
       Solvent bonding   186
       Adhesive bonding   187
   Rigidizing thermoformed parts   187
       Bonding multiple parts   187
       Foaming-in-place   188
       Fiber-reinforced structural supports   188
   Finishing and decorating thermoformed parts   190

12. Related and Competing Forming Processes 193
   Forming processes performed at lower temperatures   194
   Packaging container forming   196
   Limitations for thermoforming   197
       Injection and extrusion molding   198
       Blowmolding and rotomolding   199
       Plastisol or slush molding   199

Appendices
   A Exemplary properties of thermoforming materials   201
   B Exemplary properties of film materials   205
   C Trade names and materials manufacturers   209
   D Conversion factors   225

Index   231

Foreword

THE REMARKABLE ACHIEVEMENTS of plastics engineers, chemists, and technicians in the field of thermoforming have long been recognized by those industries that have so greatly benefited through the increased profitability engendered by attractive, convenient packaging of their products. Packaging is by far the most visible advertisement for thermoforming, but it is only one indication of the breadth of this fast-growing, ever-changing technology. Indeed, it may well be that a new era approaches in which past promises unfulfilled and radical technological innovations forthcoming will impose increasing demands on the proven ingenuities of those who have pioneered, and those who continue, the development of thermoforming technology. Again, these new technologies will be most visible in packaging as vastly improved processes for food preservation emerge in the forms of irradiation (gamma), super-pasteurization and other novel sterilizing techniques. It just may be that the long awaited fuel cell for powering automobiles may be close to reality; if so, will thermoformed automobile bodies not be far behind? New construction methods appear to be in the offing that will demand thermoformed composites providing structural strength, insulation, and weather resistance. Moreover, it would seem that in the not too distant future a group of new plastics materials will emerge that should find many applications in thermoforming; these materials will be of substantially lower cost than those of today in that their monomers are produced by direct partial air oxidation of essentially unwanted crude oil and gas oil components in a simple process of almost negligible capital and operating costs.

    A comprehensive treatise on thermoforming has long been needed; for too long this most important technology has been retained (closely held) by a relative few. This does not suggest some kind of intentional cabal, for there are numerous publications describing single processes. But an all-encompassing treatise on thermoforming technology such as this by Dr. Gruenwald is long overdue. He has ingeniously compiled an enormous amount of information and references, reformed them without loss of details, and organized the material with an "overlay" of the scientific principles involved. Hence, his book is an ideal text for all plastics engineers and technicians and for polymer chemists. Yet it is at least equally valuable to thermoformers and those who utilize thermoformed products. The generic terminology employed is especially appreciated in that students are not lost in a maze of unfamiliar, nondescriptive jargon. Dr. Gruenwald has, however, provided an invaluable delineated description of thermoforming processes replete with practical "how-to" problem-solving guides for even the most sophisticated professional in the field. And these experts will be every bit as comfortable with his terminology as novices.

Thermoforming is a highly interdisciplinary technology encompassing chemistry, physics, materials, mechanical engineering, and thermodynamics competence. Its breadth, already grandiose, may range even to metals; several alloys are now known to have remarkable plasticity, and probably many more will be discovered. Indeed, it is quite likely that glasses will be commonly thermoformed. But the great growth in thermoforming will undoubtedly continue in plastics materials. Dr. Gruenwald has indicated the desirable properties of polymerics for differing applications; thus, his text is especially useful for polymer chemists who must "tailor" plastic materials for specific groups of applications. Engineers in extruding and calendering film and sheet will benefit from the intimate relationships elucidated between processing parameters imposed upon stocks employed in thermoforming and the products thereof. Mold designers are provided with a complete guide that will enable them to avoid the less obvious pitfalls and wasted effort so often experienced in the evolution of molds for (especially) complex parts. This book is especially useful for mechanical engineers in that it delineates all of the factors, including thermodynamics, important in the design and construction of thermoforming systems. Operators will find "everything you ever wanted to know" about the technology of thermoforming, enabling them to improve productivity, minimize down-time, and greatly reduce scrap. Quite likely, Dr. Gruenwald's suggestions will lead to considerable benefits to those who read and practice by this remarkable exposition of thermoforming technology.

Dr. Gruenwald's 35 years of experience in plastics and thermoforming with Farbwerke Hoechst and General Electric, combined with the innumerable suggestions and additional information provided by other experts in this field, has resulted in a remarkably complete compilation on the subject. And his decade of teaching at Behrend College and Gannon University is reflected herein by clarity of expression, systematic organization of materials, and penetrating analyses. Combined with the complete exposition, this volume is "must" reading for all in the plastics field and most assuredly for those in thermoforming or with an interest therein. I thoroughly enjoyed reviewing the manuscript and to it assign my highest recommendation.

> ROBERT K. JORDAN
> Director—Metalliding Institute
> Director—Engineering Research Institute
> Scientist in Residence
> Gannon University
> University Square
> Erie, PA 16541

Preface

THE DIVERSITY OF equipment and the multitude of parts formable by these methods make it difficult to describe thermoforming comprehensively. Therefore, for a long time the subject of thermoforming appeared only in articles for periodicals or short chapters in handbooks. In 1987 nearly simultaneously three books came on the market solely devoted to thermoforming. The book by J. Florian, *Practical Thermoforming: Principles and Applications,* is published by Marcel Dekker, Inc. New York (1987) and James L. Throne's *Thermoforming* by Carl Hanser Verlag (1987). The latter book recently was expanded to nearly 900 pages and appeared in 1996 by Hanser/Gardner Publications, Inc. under the new title *Technology of Thermoforming*. This book contains a very large collection of data tables and charts and covers mathematically any aspect of thermoforming. Numerical examples guide the less mathematically inclined reader to their useful application. Another way to acquire knowledge about thermoforming is to attend seminars or conferences on thermoforming, which are regularly scheduled at different locations by the Society of Plastics Engineers, Inc., Brookfield, CT.

The first edition of the book *Thermoforming—A Plastics Processing Guide* appeared in 1987 by Technomic Publishing Company, Inc. (Lancaster, PA). As the subtitle indicated, this book was mainly addressed to the practitioner who is or plans to get involved in the production of thermoplastic parts outside the ubiquitous injection molding process. Every effort has been made to express material properties and behaviors in language understandable by both the specialist and the neophyte. Therefore, in general, all plastic materials are referred to by their generic names. However, a list of trade names (Appendix C) is provided for those more accustomed to using commercial terminology.

Thermoformed parts have become important in two main areas: (1) structural and functional parts and (2) low-cost, high-performance packaging applications. There is no doubt that significant advances in both areas are yet to come. This guide made an effort to describe the thermoforming processing conditions with particular attention to behavioral changes in the plastic materials. The book portrayed the types and operational sequences of standard equipment, the properties of normally used raw materials, and the peculiarities and advantages of major and minor forming processes. Also examined were cognate forming processes that compete with thermoforming.

This guide is intended to be comprehensive, a goal whose attainment is limited by the breadth of the field and the rapid evolution of new products. Several areas of intense development have been emphasized in the original book and have been further expanded and brought up to date for the second edition. These are as follows: (1) The vast expansion in regard to the availability of polymeric materials. This relates not only to the widening in copolymer and alloy compositions but also—as seen especially in polyolefins—in the increased multitude of polymerization processes. (2) The remarkable expansion of the materials' properties has further been amplified by the application of orientation and crystallization processes. (3) Where desirable properties have not become obtainable by these two means, one has found ways to obtain them by laminations or coextrusions of different materials or by surface treatments. These types of materials can be expected to gain the fastest growth rate in barrier packaging materials.

After studying the main chapters, the reader should glance at Chapter Twelve, "Related and Competing Forming Processes." At that stage, imagination plus the information derived from a few experiments should enable the reader to envision the most efficient processes and operational stages. It is hoped that the reader can join the successful businesses that use expedient production methods to manufacture thermoformed parts at favorable costs. Many of these methods were considered impossible by experts 25 years ago.

The limitations and disadvantages of thermoforming processes are explicitly discussed in this book, not to discourage, but to sharpen the technician's mind in finding ways to minimize them.

Some tables and an extended index are found on the concluding pages to help the casual reader find numerical data and additional information on subjects of interest. The reader should be cautioned that, in many cases, wide variations exist between materials supplied under the same generic names. It is therefore, always advisable to obtain data sheets from the materials' manufacturer for the specific grade of plastic to be used.

ONE
Introduction

THERMOFORMING WILL ALWAYS represent only one part of all the possibilities for shaping plastics items, but the total market is so large that if only a fraction of it goes to thermoforming, sizeable business opportunities exist. In 1997 the growth rate for the next 5 years is anticipated to maintain 4 to 6%. The Business Communications Co., Inc. (Norwalk, CT) writes that "packaging of all types is a $100 billion business in the U.S. and that the market value of plastics used to package medical, pharmaceutical and specialty health-care products was $801 million in 1995." The Rauch Guide (Impact Marketing Consultants, Manchester Center, VT) to the U.S. Plastics Industry states that "the U.S. sales of plastic resins and materials reached $150.5 billion in 1996. Among the several (five) technological advances that will contribute to future growth of plastics, the further development and use of barrier plastics for packaging applications occupies the first rank, the expansion of coextrusion film in flexible end uses were mentioned as second." Both are pointing to a respectable growth in thin film thermoforming. The third reason mentioned, "the further development of plastic blends, alloys and copolymer technology," will attain benefits also for the heavy-gauge thermoformer.

The various thermoforming processes are based on the recognition that rigid thermoplastics become pliable and stretchable when heated but will return to their original rigidity and strength when cooled. For most plastics molding processes the temperature of the material is raised until it turns into a liquid-like but highly viscous material. This makes it possible to achieve any shape as long as a suitable mold can be built for it.

In thermoforming we begin with an already preformed part, in most cases, a thermoplastic sheet or film. We limit the supply of heat so that the plastic to be formed becomes highly flexible and stretchable but still retains sufficient strength to withstand gravitational force. Only relatively weak forces are required to make the sheet conform to the surface of a mold. In most cases a fraction of the atmospheric pressure is sufficient to do all the forming. This can be readily attained by utilizing a partial vacuum between the sheet and the mold or by pressurizing an air chamber to which the sheet has been sealed. In either case the air passes through small air holes distributed over the mold surface. The most widely used thermoforming processes are commonly designated as vacuum forming. There are many thin-walled articles that cannot be produced profitably by any other way than thermoforming.

Two further steps are required to complete production of a useful part. The formed sheet must be cooled to become rigid again, and, finally, the excess material around the circumference must be trimmed off. The cooling and heating processes represent the most time-consuming steps and must, therefore, be painstakingly arranged.

The first important plastic utilized in the thermoforming process—as in injection molding—was celluloid, a highly flammable cellulose nitrate-based plastic, which many years ago was made obsolete by more suitable alternatives. The twin-sheet forming process was used, in which two sheets of celluloid were mounted between two complementing or identical mold halves. These sheets were then heated and expanded into the molds by pressurized steam. The objects produced consisted primarily of children's toys and small low-cost containers.

Unlike all other plastic forming processes, thermoforming can be adapted to so many variations that it becomes difficult to cover all aspects in a single text. If all the practiced process combinations were described, too many details would have to be repeated.

In most other processes for forming plastic parts, the possibilities for variations of the part's wall thickness lie within quite narrow ranges. This is especially true for injection molding and blow molding. For thermoforming no such limitations exist. A thin film, 0.001" thick, as well as a sheet 1" thick can be thermoformed. Because the time frames for heating and cooling in these instances can vary more than 1000-fold, it should become obvious that many of the other processing details would vary also. Therefore, thermoforming has become divided into thin-gauge (less than approximately 1/16") and heavy-gauge (greater than approximately 1/8") thermoforming processes. For practical purposes this division should rather be made between processes utilizing a film fed to the former continuously from a roll or an extruder on one hand and a process in which a precut flat sheet is individually clamped in a frame and formed. The resulting products similarly can be parted into mass-produced, nearly no-cost, thin wall, short-lived packagings, and the structurally strong, durable, utilizable objects or parts of appliances.

TWO
Heating of the Plastic

THE HEATING PROCESS can be one of the most time-consuming steps in thermoforming. Although the scientist isolates heat transfer into three distinct phenomena—conduction, convection, and radiation—in practice they will mostly be concurrent. The differentiation between thin film and heavy sheet thermoforming is in no other processing step more evident than during the heating of the plastic. The peculiar physical properties of plastics dictate the appropriate selection to obtain the most efficient heating method.

In June 1995 a 42-page report on "Heating Technologies for Thermoforming" was issued by the Electric Power Research Institute (Center for Materials Fabrication, Columbus, OH). It presents a comprehensive overview of present infrared heating systems.

Means of conveying heat to the plastic

Contact with a uniformly hot metal plate, which is *conduction heating,* is preferred for the heating of very thin films. Heating is accomplished in a fraction of a second, which is of great importance for forming biaxially oriented films. This method is used especially for the mass production of smaller, thin-walled articles and is called *trapped sheet heating* because the sheet is held to a hot plate by a slight vacuum or by applying a slight air pressure from the outside. Such heating plates consist mainly of aluminum and have a nonstick fluorocarbon polymer coating. They contain a number of small holes and are generally located directly above the mold but could also be used as sandwich heaters. Conduction heating provides the most uniform and consistent temperature distribution in film or sheets and by far the best energy efficiency. Where sandwich contact heaters are used, instant heating takes place and gauge fluctuations of the material will have no effect on the forming process.

Due to the low infrared absorption of polyolefins and especially the fluoropolymers only a small amount of radiant energy would be utilized if these thin films were heated by radiation.

Steam heat represents an ideal *convection heat* source both in regard to heat output and uniform temperature distribution. Its convection heat transfer coefficient is 1000 times greater than that of air. However, the application of steam as a heat source is limited to cases where the generated condensed water poses no problem and where this very limited temperature range (212°F) can be tolerated. Presently, steam heat dominates the production of polystyrene foam parts and foam boards. Steam heat had its day at the beginning of thermoforming, when toys etc. were fabricated from celluloid using the twin-sheet steam injection process that provided heat and pressure simultaneously. It is no longer utilized for thermoforming, because most of today's plastics require higher temperatures.

Present-day convection heating of plastics is accomplished in two different ways. Heating the plastic by submerging it in a *heat transfer liquid* is very efficient due to the 100 times higher convection heat transfer coefficient of liquids versus air. However, its use is restricted to heavy-gauge parts because the removal of the liquid poses problems. Its main advantage is that here also a very uniform heat distribution can be obtained in a relatively short time.

Air convection ovens are used extensively for heavy sheets because they provide quite uniform heating. Such ovens, when set at lower temperatures, are also suitable for drying sheets that have been exposed to moisture. If the temperature of these ovens is kept at the forming temperature of the sheet, heating times are long, but a wide margin of safety in regard to heating time variations is obtained. Because the effective convection heat transfer coefficient from turbulent air to solid plastic is low (0.002 to 0.010 W/sq in. × °F) and the heat conductivity of the plastic is also very low, heating times are long and increase with the square of the sheet thickness.

With all these methods, a considerable preheating time is required for the equipment before the heat can be transferred to the first piece of plastic. On the other hand, some infrared radiation heating equipment can supply instant heat and, therefore, needs only to be turned on for the seconds the plastic sheet is heated.

Microwave heating methods have found only few applications in thermoforming due to the great expense of such equipment and the difficulty of obtaining satisfactory heat distribution over large areas. Besides, dielectric heating is unsuitable for most thermoplastics. Although heating times can be cut to 10%, the extended cooling time requirements would still remain unchanged. Internal heating methods are applicable to forming processes where only localized heat is required, e.g., edge forming of materials having high dielectric loss factors, such as polyvinyl chloride sheets.

Physics of radiation heating

Radiation heating occupies the largest share of plastics heating for thermoforming. The energy density, which can be transferred by infrared radiation from the heat source to the plastic sheet, depends on many variables. The higher the radiation energy density—usually expressed in watts per square inch—the less heating time will be required. The limit for heat flux for thin films is approximately 0.02 to 0.05 W/sq in. × °F to obtain short heat-up times. For films the heater times increase proportionally with thickness. Due to the low thermal conductivity of the plastic the surface temperature of thick sheets may quickly surpass the thermal stability of the plastic, making it necessary to lower the radiation energy. This can be achieved on roll-fed thermoformers by lowering the wattage to the second heater bank or on individually heated sheet formers by reducing the tubular quartz heater output in one or two steps. Slow-response heaters can be kept throughout the whole heating cycle only at the lower energy setting.

The surface temperature of practical radiant heaters ranges between 600 and 1800°F. Figure 2.1 illustrates the relative emissive power radiated by an ideal black body surface with an emissivity of 1. In Figure 2.1 the areas enclosed between the baseline and the respective temperature curves express the amount of energy emitted by an ideal infrared heater. The heat energy of available heaters may amount to only 75 to 95% of that amount due to lower emissivities.

As a general rule for assessing *emissivity* values, it can be stated that all heater surfaces have a high value of approximately 0.95. This includes, besides oxidized stainless steels, other oxidized metal surfaces, such as iron, nichrome wire, and galvanized steel. Glass and ceramic surfaces, including white glazed porcelain, exhibit equally high values. On the other hand, highly polished metal surfaces are usually at or below 0.05 and can, therefore, advantageously be utilized as reflectors of heat radiation. All metals, including gold, silver, copper, aluminum, nickel, and zinc, fall into this group. The difference lies in their capability of retaining the shiny polished surface. *Absorptivity* should theoretically be identical to the emissivity, but it relates in thermoforming processes to the cold materials exposed to radiation. Therefore, different temperatures and wavelength ranges dominate. Most nonmetallic materials have a value of 0.90, and this applies to wood, glass, plastics, textiles, and paints. There is only a small difference between a black and a white paint or plastic because the binder resin or polymer controls the absorption.

It can be seen that at higher temperatures the bulk of radiation occurs at lower wavelengths, 3 to 4 microns. By contrast, at lower temperatures the considerably lower radiation energy is spread out over a broader and

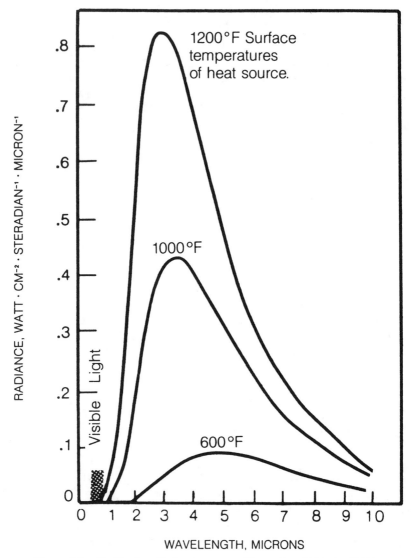

Figure 2.1. Blackbody radiation (courtesy of Ircon, Inc., Niles, IL 60648).

higher wavelength range, 4 to 7 microns. This is important when thermoforming thin sheets or films because each plastic material absorbs infrared radiation in distinct regions. Only absorbed radiation is utilized for heating the plastic directly. The remainder will just heat up the oven enclosure and thus indirectly heat the sheet or will be lost to the outside.

The portion of radiation that is reflected by the sheet is insignificant and nearly independent of the wavelength and naturally independent of sheet thickness. Unless one wishes to thermoform a metallized plastic sheet, reflection losses, which normally account to no more than 4%, can be ignored. Energy losses due to transparency in certain wavelength regions, however, can be significant for very thin films. From transparency in the visible region one cannot conclude whether the sheet will be opaque or transparent at the frequency of maximum heat radiation. As seen in the graphs of Figure 2.2 (pages 8 and 9), transparency increases markedly as the thickness of the film decreases. Sheets of 1/8" thickness will absorb practically all infrared radiation. On the other hand, thin films of less than 5 mils will be inefficiently heated when irradiated in certain infrared regions, e.g., fluorocarbon polymer films are transparent up to 7.5 microns. This is similarly true for thin polyolefin (polyethylene, polypropylene) films. In these cases radiation passes through the film and heats the surrounding chamber, which then indirectly heats the film at a higher wavelength. But the rise in heater temperature results in higher energy losses. On the other hand, a radiation in a less suitable frequency range may be applied to advantage by letting the energy penetrate deeper into the sheet material, thus avoiding scorching the sheet surface and making it possible to irradiate at a higher wattage.

The relationship between heater surface temperature and heat energy transfer can be estimated using the Stefan-Boltzmann relationship:

$$E = \sigma\varepsilon \left[\left(\frac{T_{heater}}{100}\right)^4 - \left(\frac{T_{plastic}}{100}\right)^4 \right]$$

E = Radiation energy per unit area in $\frac{Btu}{sq\ ft \cdot hr}$ or in $\frac{W}{sq\ in.}$

σ = Stefan-Boltzmann constant $0.1713\ \frac{Btu}{sq\ ft \cdot hr}$

ε = Emissivity of heater and plastic surfaces (assumed to be 0.9)

T = Temperatures in °R (Rankine) equal to °F + 459.6

Example: Determine the emitted energy E when heater surface temperature is 1000°F (1460°R) or 1500°F (1960°R) and the plastic surface at 100°F (560°R).

$$E_{1000°F} = 0.1713 \times 0.9 \left[\left(\frac{1460}{100}\right)^4 - \left(\frac{560}{100}\right)^4 \right]$$

$$= 6850\ \frac{Btu}{sq\ ft \cdot hr} = 14\ \frac{W}{sq\ in.}$$

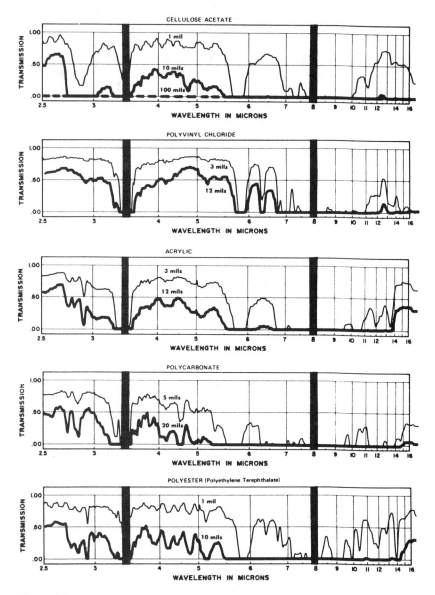

Figure 2.2. Transmission spectra of several commercial plastics (courtesy of Ircon, Inc., Niles, IL 60648).

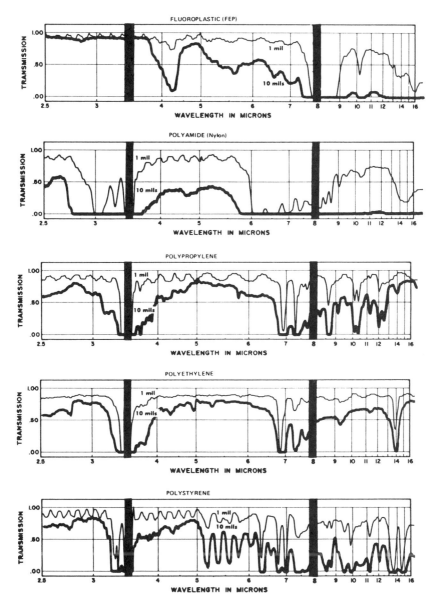

Figure 2.2 (continued). Transmission spectra of several commercial plastics (courtesy of Ircon, Inc., Niles, IL 60648).

and

$$E_{1500°F} = 0.1713 \times 0.9 \left[\left(\frac{1960}{100}\right)^4 - \left(\frac{560}{100}\right)^4 \right]$$

$$= 22600 \ \frac{\text{Btu}}{\text{sq ft} \cdot \text{hr}} = 46 \frac{\text{W}}{\text{sq in.}}$$

If emissivity would be lowered to only 0.5 at a constant wattage of 14 W/sq in., the heater surface temperature would rise to 1225°F. Although the radiation in each case is spread out over a wide band of wavelengths (see Figure 2.1, page 6), the bulk of energy would be radiated at:

- 1000°F at approximately 3.5 micron
- 1225°F at approximately 3.0 micron
- 1500°F at approximately 2.5 micron

Energy emitted from a heater surface rises drastically with increasing temperature. The primary emitter surfaces of near-infrared heaters are relatively small (tungsten wires), but the bulk of heat will be radiated by the indirectly heated quartz tube. Still heaters are spaced at wide intervals and must, therefore, be increasingly distanced from the plastic. Some relative motion or oscillation of either heater or plastic is advantageous for streak-free heating. Because radiation from the wires travels in all directions, reflectors behind these heater elements are essential.

Heaters with totally enclosed resistance wires or bands occupy a larger area but still require the reflection of the radiation emitted toward the opposite direction. This type of heater can be seen in the photographs of large, automatic thermoformers (Figures 2.3 and 7.15, page 91). Finally, surface ceramic or metallic heaters are spread out over the entire area of the heater assembly. These units are backed by a thermal insulation to reduce energy losses (Figure 2.4).

The importance of *high surface emissivities* for the efficiencies of resistance heaters is frequently overemphasized. One hundred percent of the electric power will be converted to heat in all cases. Most of the heat absorbed by the plastic will be due to radiation, and most of the heat lost to the environment will be due to air convection. Reduced emissivity will raise the temperature of the heater, thus shifting the radiation toward the near-infrared range. Only a few thin, clear films, such as polyolefins and fluorocarbon polymers, tend to absorb this kind of radiation less efficiently. For such pigmented or heavier films and sheets this shift is actually an advantage, because the shorter wavelength radiation will penetrate and generate heat deeper in the plastic.

The concentrations of pigments in colored sheets are generally low

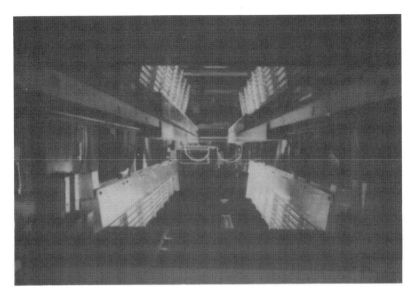

Figure 2.3. Infrared heater. Tubular resistance elements arranged on upper and lower split clam shell ovens for quick access and safety.

Figure 2.4. Ceramic infrared radiators. Array of heaters mounted on top of oven (courtesy of Infrared Internationale, Ltd., Ballydehob, Rep. Ireland).

enough so that no significant variations will occur. Carbon black may be one exception and may increase absorption in thin films. White pigmented sheets may show somewhat higher reflection losses, but only in the near-infrared region.

Although conventional hot air convection ovens might have to be modified to obtain better temperature uniformity over the entire sheet, the problems of obtaining consistently *homogeneous heat distribution* when using radiant heat are enormous. This is especially true if the sheets suffer from excessive gauge variations. There is a great temperature differential between the heaters, the frames, and the various clamping devices. In some cases the latter two must be water cooled; in others they are heated or at least preheated. Nevertheless, every square inch of the plastic should ideally have exactly the same temperature at the end of the heating cycle unless special heating schemes are desired. The difficulty of obtaining a low temperature gradient also throughout the thickness of the plastic sheet will later be covered.

The sketches in Figure 2.5 illustrate how to compensate for heat losses occurring at the outside edges. It can readily be seen that less radiation strikes the plastic on the outskirts of the sheet especially when the distance between the heater and the plastic is great [arrow points hitting each square of the sheet in Figure 2.5(a)]. Therefore, either additional heaters must be mounted at the outer edges and especially outer corners [Figure 2.5(b)] or the inside heaters must be spaced farther apart [Figure 2.5(c)] or energized at lower wattage.

To appraise these inevitable energy losses at the corners and edges, one must consider the *view factor* of an installation. In this case the surface area of the heater is identical to that of the plastic sheet. The parameters, which can be obtained by dividing the sides by the distance between the heater and the sheet (M = heater width/distance and N = heater length/distance), just must be entered into the graph in Figure 2.6 to read the resulting view factor on the y-axis of the graph. This value is identical to the fraction of the total *average* radiant heat that strikes the plastic.

Example: Assuming that the dimensions from Figure 2.7 for the square sides are 25" and the distance between the heater and sheet 5", the values for M and N become 25/5 = 5. The resulting view factor is, therefore, $F = 0.7$, meaning that 30% of the energy is directed outside of the plastic sheet.

In practice the results are much worse because this reduced amount of energy is not distributed evenly over the whole surface. Figure 2.7 illustrates this uneven distribution, in which case only the most central part receives the full energy density (100% of the heater's W/sq in. rating) from the emitter.

Reducing the distance between heater and the plastic and positioning

Physics of radiation heating 13

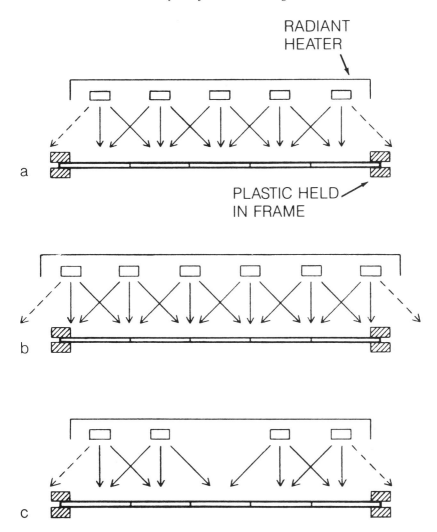

Figure 2.5. Heater arrangements for obtaining good heat distribution.

the reflectors at the heater edges will lessen these edge losses. Sandwich heaters (simultaneous exposure of the plastic to radiation from both sides) may also mitigate both radiation and convection losses. When sandwich heaters are used, the lower heater banks should have a wire grid above the surface to prevent the sheet from contacting the hot heater elements in case of excessive sagging or tearing.

One should intentionally apply excessive heat to the edge areas of the sheet to better utilize the material by drawing more of it away from the skirt area, which in most cases must be trimmed off and is many times discarded. On the other hand, especially during production of highly

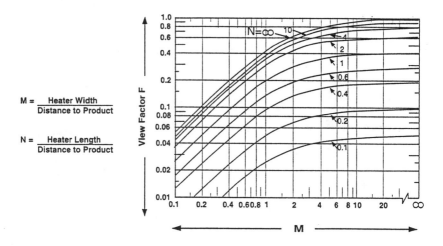

$M = \dfrac{\text{Heater Width}}{\text{Distance to Product}}$

$N = \dfrac{\text{Heater Length}}{\text{Distance to Product}}$

Figure 2.6. View factor diagram (courtesy of Electric Power Research Institute, Center for Materials Fabrication, Columbus, OH 43215).

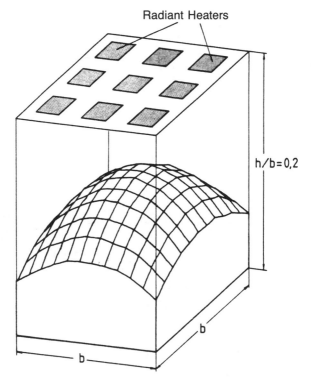

Figure 2.7. Radiant flux received on a 25-in. wide square plastic sheet from a 5-in. distanced uniformly powered heater surface of equal size (courtesy of D. Weinard, Doctoral Dissertation, Institut für Kunststoffverarbeitung, Aachen, Germany).

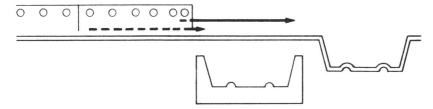

Figure 2.8. Roll-fed automatic thermoformer. Length of solid arrows indicates time that parts of film are exposed to cooling prior to forming.

unbalanced parts, certain areas of the plastic sheet must be kept at a lower temperature to diminish the reduction in material thickness in those areas. By shading, i.e., using wire mesh screens of various densities and sizes (supported on metal grates under or above the heaters), one can quickly establish the most effective temperature profile. For long production runs it becomes more practical to obtain such zoning by using—instead of screens—individually controlled heater banks, which allow the adjustment of heater wattage for specific areas.

When forming thin sheets or films—especially with roll-fed automatic thermoformers—one should keep in mind that *the leading edge of the material,* which leaves the oven first, loses more heat than the trailing edge. This can be alleviated by positioning the heater elements closer together at the film withdrawal end of the final heating station and by outfitting the heater edge with baffles to reduce heat losses (Figure 2.8).

Thermal properties of plastics

The following material properties must be taken into consideration when selecting a heating method and the rating of heating equipment.

Plastics are poor heat conductors (Table 2.1). Therefore, thick sheets will require a disproportionately longer heating time. Heating the sheet from both sides (sandwich heating) will help to decrease this time. In some cases heating time can be reduced by preheating the material and keeping it at an intermediate temperature. This is, however, seldom done for materials under 1/4" thickness. One vendor (Figure 2.9) supplies for roll-fed thermoformers a preheating station that slowly preheats the sheet, looped through a convection oven, to a very uniform temperature within $\Delta 3°F$ before heating it quickly by radiation to the forming temperature. This procedure eliminates many of the problems encountered when forming polypropylene, e.g., expansion of the sheet, sagging, and nonuniform temperature distribution.

In addition, the heat content of plastics is high compared with any other material, water being the main exception. Approximately 60 to 100 Btu

TABLE 2.1
Thermal properties* of materials.

Material	Specific gravity $\frac{g}{cm^3}$	Bulk density $\frac{lb}{cu\ ft}$	Specific heat $\frac{Btu}{lb\ °F}$	Heat of fusion $\frac{Btu}{lb}$	Thermal conductivity $\frac{Btu\ ft}{sq\ ft\ hr\ °F}$	Linear coefficient of thermal expansion $\frac{in.}{in.\ °F}\ 10^{-5}$
Air	0.0012	0.07	0.24		0.014	
Water	1	62.4	1	144	0.343	
Ice	0.92	56	0.5	144	1.26	2.8
Wool felt		5.6	0.41		0.021	
Syntactic foam			0.50		0.076	1.7
Hard wood, parallel to grain	0.7	45	0.4		0.08	0.3
Hard wood, across grain	0.7	45	0.4		0.05	3
Polytetrafluoro-ethylene	2.13	133	0.25		0.146	4
Phenolics	1.5	95	0.3		0.2	3–5
Epoxies	1.6–2.1	100–130	0.3		0.1–0.8	1.5–2.8
Polyethylene, high density	0.96	59.9	0.37	55	0.28	7
Polystyrene	1.05	65.5	0.32		0.07	3–4
Acrylics	1.19	74.3	0.35		0.108	3.5
Polycarbonate	1.20	74.9	0.30		0.112	3.7
Graphite	1.5	93	0.20		87	0.44
Glass	2.5	160	0.2		0.59	0.5
Quartz, fused	2.8	165	0.2	102	0.8	0.029
Aluminum	2.7	165	0.23	171	90	1.35
Steel	7.8	490	0.10	117	27	0.84
Copper	8.8	550	0.092	88	227	0.92

*Refer to Appendix D for the conversion factors for listed values.

Figure 2.9. Roll-fed material preheating station (courtesy of Paul Kiefel GmbH, D-83395 Freilassing, Germany).

are required to heat 1 lb of plastic over a temperature span of Δ200°F. Crystalline plastics will require an additional amount of heat, approximately 20 to 60 Btu, to render them thermoformable. For assessing heat requirements one can easily estimate a value by multiplying the following factors:

Heat Requirement = Length × Width × Thickness × Density

× (Specific Heat × Temp. Difference + Heat of Fusion)

For example, the requirement for continuous thermoformer heating of a 0.010″ thick sheet of low-density polyethylene (specific heat of 0.55 Btu/lb·°F and a heat of fusion of 33 Btu/lb) with a sheet width of 42″ and a sheet speed of 100 ft/min from 70 to 350°F will be about:

$$1200 \frac{\text{inch}}{\text{min}} \times 42 \text{ inch} \times 0.01 \text{ inch} \times 0.033 \frac{\text{lb}}{\text{cu inch}}$$

$$\times \left(0.55 \frac{\text{Btu}}{\text{lb °F}} \times \Delta 280°\text{F} + 33 \frac{\text{Btu}}{\text{lb}} \right) = 3100 \frac{\text{Btu}}{\text{min}}$$

which is equal to 55 kilowatts.

This relationship becomes simpler for noncrystalline plastics due to the omission of the heat of fusion. When the working parameters have been well established for one such product, only those multipliers that have changed must be considered for any replacement material, to obtain the new heat requirements.

The coefficient of thermal expansion might also have to be considered by the manufacturer of thermoforming equipment. Again, polypropylene can serve as an important example. During heating the sheet is expanding to such an extent—partly due to the melting of the crystalline portion—that it would contact the lower heater banks if the parallel clamping rails would not be led at a diverging path at those regions. These difficulties might become exasperated by the shrinkage that might occur in the longitudinal direction if the sheet has been pulled excessively during extrusion.

The necessity for observing the upper boundary of the thermal stability of the plastic material will be emphasized later. The surface appearance of plastics, such as gloss or mat finish and especially imprinted surface texture, can become affected at temperatures that are considerably lower than the thermal stability of the plastic itself. In these cases when using sandwich heaters, lowering the heater output at the appearance side of the sheet is recommended.

Heating equipment for plastic sheets

Convection ovens were originally the most common devices used to heat plastic sheets for thermoforming. They are still the preferred way for heating very thick sheets and for distributing heat uniformly where free-forming techniques are utilized.

The heat can be supplied by gas flames or by electric resistance units. Forced circulation of air and baffling to equalize the air flow at around 200 feet per minute are crucial to obtain temperature uniformity and adequate heat transfer. Good thermal insulation of the oven walls and the strategical position and size of entrance and exit doors increase energy efficiency. A typical forced-circulation air oven is shown in Figure 2.10 and a practical means for hanging sheets on trolley tracks in Figure 2.11. When sheets are heated horizontally to better utilize the heater space, they are supported in trays. These should be lined to protect the surface of the plastic. A polytetrafluorocarbon-coated glass fiber cloth lining is ideal.

Automatic temperature regulators must be provided to keep air temperature fluctuations as low as possible and temperature recorders will facilitate better supervision. Additional limit controls should be set to avoid fires or melting of the plastics in case of oven malfunctioning.

The temperature of the oven should be set at the forming temperature of the plastic. This way, even sheets of variable thickness can be left in the oven for irregular time intervals until the forming equipment or manpower is ready for continued processing. At higher and steadier production rates, the oven temperature will probably be set higher, but the sheets must be withdrawn of the oven according to an established time schedule to prevent tearing or thermal degradation.

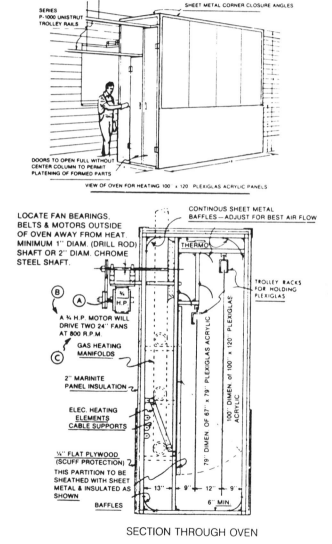

Figure 2.10. Typical forced-circulation air oven (courtesy of Rohm and Haas Co., Philadelphia, PA 19105).

20 Heating of the Plastic

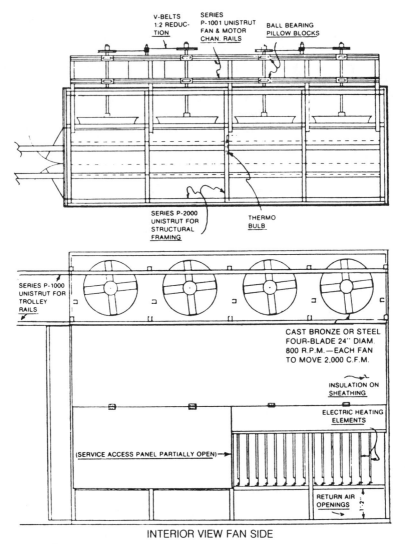

Figure 2.10 (continued). Typical forced-circulation air oven (courtesy of Rohm and Haas Co., Philadelphia, PA 19105).

Infrared radiant heating is—besides dielectric heating, oil submersion heating, and contact heating (all three very much limited in their application)—the fastest way of heating plastic sheet or film to thermoforming temperature. Due to the radiation sensitivity of plastics, which vary with their chemical composition, direct radiation heating is limited in most cases to the outer 10 mils of the sheet. The temperature at the center (95 mils of a 1/8″ thick sheet) will be raised only by the conduction from the outer skin. Because plastics are poor thermal conductors (see Table

2.1, page 16), the high thermal gradient sets limits to the energy density (watts per square inch) of the heaters and will affect total heating time. Still, heating times for infrared radiant heating of 1/8" thick sheets can generally be held to 1 minute at 10 W/sq in. energy input.

Although heater energy densities may vary with equipment, also differences must be considered in regard to materials used. High-temperature plastics, such as polysulfones, polycarbonates, and polyesters, are the highest, with approximately 30 W/sq in. The cellulosics, styrene- and vinyl-polymers and -copolymers are the lowest at approximately 15 W/sq in. The crystalline polymers, such as the polyethylenes and polypropylene, are in the region between 20 and 30 W/sq in.

Using the formula given on previous pages the following calculations can be made:

Example: One square inch of 0.125" thick acrylate sheet of 0.043 lb/cu in. density and 0.35 Btu/lb·°F specific heat and a temperature rise of $\Delta 270°F$ will require theoretically:

$$1 \text{ in.} \times 1 \text{ in.} \times 0.125 \text{ in.} \times 0.043 \frac{\text{lbs}}{\text{cu in.}}$$

$$\times 0.35 \frac{\text{Btu}}{\text{lb} \times °F} \times \Delta 270°F = 0.508 \text{ Btu}$$

or 36 seconds at 15 W/sq in.

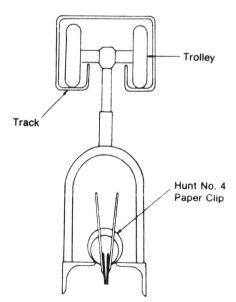

Figure 2.11. Sheet hanging device (courtesy of Rohm and Haas Co., Philadelphia, PA 19105).

Thin films can be heated at higher energy densities in a considerably shorter time. Therefore, depending on application, infrared radiant heaters for thermoforming are supplied with heat outputs of 5 to 50 W/sq in.

Both thin films and heavy sheets are heated 20 to 50°F higher at their surface than required for the forming process. The thin films lose heat in fractions of a second, the time necessary to move the film from the oven to the forming station. Heavy sheets can be held in transit or in the forming station for up to a few minutes to transfer some of the excess surface heat to the inside. This minimizes "mark-off" imperfections due to mold surface roughness, dirt particles, or glove marks when hand forming is practiced. Figure 2.12 illustrates these changes in temperature profiles.

The lowest cost *gas-fired infrared heaters* are surface combustion burners in which the gas-air mixture is burned at the surface of either a stainless steel or ceramic porous structure. Under these circumstances the plastic must be kept at greater distances than in heaters where the flame is hidden in hollow ceramic structures. Safest are gas heaters where combustion occurs in a metallic tube with no openings inside the heating chamber. This also minimizes drafts.

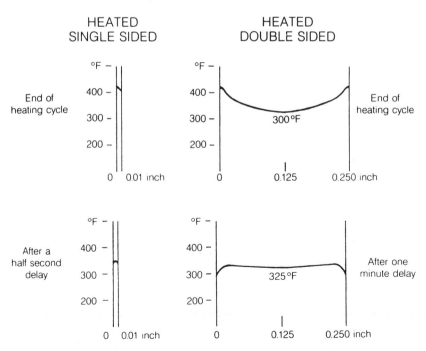

Figure 2.12. Temperature profile of radiation heated plastic.

Heating equipment for plastic sheets 23

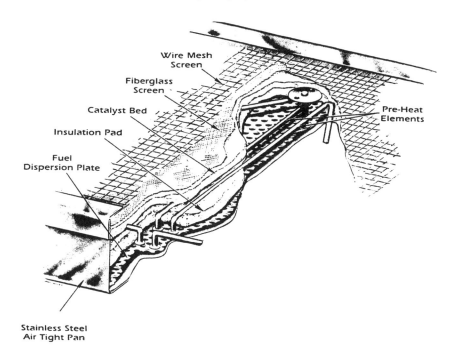

Figure 2.13. Cross section of catalytic heater (courtesy of Vulcan Catalytic Systems, Portsmouth, RI 02871).

To take advantage of the lower energy cost for natural gas without having to take a risk of fire, the above listed heaters have given way to *catalytic gas heaters* with flameless combustion. As shown in a cross-sectional cut of such a heater (Figure 2.13), natural gas or propane enters the back of the gastight heater pan and is distributed through a preheated catalyst bed consisting of platinum-coated fiber mesh. Oxygen is provided from the air passing over the front side of the heater. The overall energy efficiency of 72% is affected mainly due to the inevitable air convection. These catalytic gas heaters are, however, more expansive, need electrical preheating, and are more complicated for the adjustment of their output. The use of volumetric flowmeters should be most suitable to obtain reliable adjustments, although generally just pressure settings are used. The surface temperature of catalytic heater panels is only 650 to 800°F, thus limiting watt density to 6 to 12 W/sq in. This energy density may not be sufficient for heating engineering thermoplastics, which require forming temperatures above 400°F. There still remains a remote possibility of fire when a sheet drops onto the heater surface choking the access of sufficient air for complete combustion.

Electrically powered infrared heaters are available in a wide range of designs. In order of decreasing radiant surface temperatures (i.e., increasing wavelength of the energy emitter) they are:

(1) *Tungsten wire filament heaters* in quartz tubes and tungsten wire filament glass lamps, including halogen lamps (the filament temperature can reach 4000°F, but face temperatures are kept at 1100 to 1800°F). Because these tubes do not occupy the whole area of the heater enclosures, they either feature a (gold-plated) internal reflector or they must be provided with efficient reflectors behind them. To overcome the uneven heating pattern, the distance between heaters and sheet must be increased unless either one of them can be held under oscillating motion. One of the advantages of these types of heaters is that their warm-up time is just a fraction of a second. Their disadvantage lies in their fragility.

For heavy sheet applications halogen lamps are now being introduced in Europe to take advantage of cutting the heat-up time in half. The halogen heat elements are packed closely so they can radiate at up to 50 W/sq in. power. The radiated energy at the high temperature of 2000°F is concentrated at 1 micron infrared wavelength and is therefore absorbed less at the sheet surface but penetrates approximately 1/32" deep. Because heat comes on instantly and is required only for a minute, albeit at a high power, the electrical circuits must be provided for high amperage, posing an installation penalty.

Quartz tubes with nichrome heating wire coils are limited to a maximum operating temperature of approximately 2000°F. To reach full power—also at up to 50 W/sq in.—a warm-up time of 1 to 3 minutes is required.

(2) *Electrical radiant panel heaters* are also sold in a great variety of designs in regard to their emitter surfaces. Basically, they all consist of nichrome wire coils that are embedded closely in grooves of a refractory plate and heated to approximately 2000°F wire temperature. To minimize heat loss, their back plates are insulated and enclosed in a metal frame. The more expensive ones feature quartz plates in the front; others utilize just quartz cloth, the glass plates Ceran® made by Schott or Neoceram® made by Nippon Glass or coated stainless steel and aluminum. Their advantage lies in their sturdy construction, good efficiency, and easy cleanability. Their heat output is quite uniform with 10 to 30 W/sq in. at a range of 1100°F up to an excessively high 1800°F emitter temperature. But they require long heat-up times (10 minutes) and are less suitable where intricate heat patterns are required.

(3) *Infrared ceramic heaters* come in smaller standardized sizes, requiring the assembly of large numbers of them for a typical heater (see

Figure 2.4). The flat plate heaters have an integral insulator at their back side and should be arranged to occupy the whole surface of the heater assembly. They are the ones that are most interchangeable with radiant panel heaters. The concave heaters radiate also to the rear side and must therefore be spaced dispersed in front of a reflecting surface. They require a greater distance to the plastic sheet. Because the heating wires are tightly imbedded in a highly emitting white ceramic body, they are protected from oxidation and can deliver up to 40 W/sq in. at surface temperatures up to 1400°F. Their heat-up time is approximately 5 minutes. Intricate heating patterns can be established either by means of individual heater embedded thermocouples or by a computer program utilizing a scanning infrared sensor. Variant heaters with a coating providing a color indication for judging proper heating operation are also available.

(4) *Tubular metal-sheathed heaters* (usually called calrod heaters) dominated, at one time, heater construction due to their ruggedness, low cost, and long service life. They consist of nichrome wires or bands coiled helically or into flat strips. Magnesium oxide, which is a good thermal conductor, serves as an electrical insulator to prevent contact with the stainless steel sheath. Heat-up time is very long, and heat output is only moderate, not easily controllable, and deteriorating with time. Their low efficiency is mainly based on the required extended distances between heaters and the plastic.

Misconceptions also persist regarding an alleged drop in energy efficiency with radiant heat elements due to the fact that often heater times must be increased with passage of time. The emitted energy of quartz heating elements remains fairly constant. If less than the original heat strikes the plastic sheet, it is usually due to the reflectors having accumulated dirt as production progresses. On the other hand, metal-sheathed tube and plate heaters emit less heat as the nichrome wire gradually oxidizes. The wire, now with a smaller diameter, will conduct less electricity, and, therefore, will use and deliver lower wattage. A variable voltage transformer or solid-state heater controls can compensate for this apparent loss in efficiency.

In electrical resistance heating the electrical energy is always totally converted into heat energy. The wavelength of the emitted energy correlates inversely with the heater surface temperature. As the temperature goes up, the radiated wavelength goes down. But the energy density increases much more steeply (see Figure 2.1). Losses are mostly traceable to excessive air convection, escape of radiation, contamination of heater or reflector surfaces, and damages to the oven insulation. Changes in these conditions should be watched for after prolonged use.

The small-diameter, coiled nichrome wire heaters have the shortest life span, and the larger panel or plate heaters have the longest. The life expectancy will drop somewhat with increasing temperature, but frequent and deep temperature cycling may also affect heater life, mainly on account of movements caused by differences in the thermal expansion of resistance elements, ceramic insulation, and metal sheathing. Therefore, a variable transformer control or an SCR control improves heater performance. Other fast-cycling, time-proportioning controls, especially in conjunction with quick-acting mercury displacement relays, will also perform well. If low-cost, slow acting on-off controls are to be used, it may be advantageous to leave one heater bank or station on full time and only cycle other units, such as the bottom bank.

The physical arrangements of heaters can vary widely with each installation. For heating large and heavy parts, the sheets might be supported on hangers and conveyed through bulky convection ovens as illustrated in Figure 2.9. Smaller, individual sheets are usually first supported in a frame and then either transferred into the oven, or heaters are passed over (and under) the sheet. Side-by-side thermoformers with only one heating stage (shuttle thermoformer, Figure 7.4, page 83) permit full-time utilization of the heaters. Roll-fed and automatic thermoformers are provided with long heat tunnels (Figures 2.3 and 7.13, page 89) for optimal heat utilization. This ensures that the heating cycle of the thermoforming process will not become the limiting factor. Depending on the length of the stroke, the heater can be split in two or three zones of identical length, each of which could be individually powered. All of these thermoformers are equipped with special safety mechanisms (e.g., clam shell construction or oven rollout) to prevent overheating or flame-up in case the machine must be suddenly stopped (Figure 7.15, page 91). Heaters for automatic thermoformers must be adjusted when the indexing stroke is changed, so that total effective heater length will become exactly a multiple of sheet stroke length.

Sophisticated heaters employ schemes to cut down heating times by moving heater and sheet in unison over the mold or by starting the heating cycle with a surge of power. For single-unit heaters this is possible only with quartz heating elements with tungsten filament heaters that will instantly emit full power and, therefore, need only be energized while the material is heated. Ceramic heating elements have a greater heat lag. Therefore, more time must be allotted for changes in temperature settings until the heater output corresponds to the new adjustment. At the other extreme, convection ovens may need hours of warm-up time until they reach temperature stability.

Both the heater and the forming area should be free of uncontrollable air drafts because it is difficult to enclose them hermetically. When heated, many plastics will give off plasticizers or odorous—though not necessarily toxic—fumes, which may be considered a nuisance. The areas where they escape can sometimes be spotted from the rising plume. It is advisable to install at those points a very low-powered exhaust hood sufficient to skim off the fumes without causing drafts.

Judging the correct temperature of the heated sheet

In Table 2.2, the recommended sheet temperatures and the maximum mold temperatures for thermoforming of various plastic materials are listed. For two-dimensional forming (bending) the lower range would most likely suffice, whereas for other forming processes, where sheets are highly stretched, the upper temperature range should be considered.

One empirical way of establishing the most practical spot within the thermoforming window is to start out with heating a sheet to its maximum temperature. This upper limit becomes recognizable either by excessive sag, making it nearly impossible to transfer the sheet to the forming station or by noticing sheet discoloration or loss of optical clarity (blister formation or whitening). Most likely, the formed part will thin out too much and will tend to show webbing when formed over a male mold. By reducing the heating time in increments for successive forming steps—all other parameters remaining unchanged—the obtained part will at one point indicate at inside corners or other detailed areas that the sheet has not contacted 100% of the mold's surface. This temperature is indicative that the lower temperature limit of the window has been surpassed. Best material distributions are generally obtained with temperatures comfortably above the lower limit.

It is impossible to establish the absolute sheet temperature accurately with contact thermometers or thermocouples. Only radiation pyrometry and infrared sensors represent accurate methods for determining the actual surface temperature of plastic sheets and films. However, it is necessary that the instrument is calibrated for the emissivity of the material and that no stray radiation from reflections of the heaters impinges on it. It must further be cautioned that the temperature reading obtained represents only the surface temperature. Heavier gauge sheets, especially if heated with high-energy density, might have a considerably lower temperature in their center. Under these circumstances, after removing the sheet from the heater, the average sheet temperature will appear at the end of a sharp drop in temperature (a few seconds to 1 minute). As an optical device, the pyrometer can be mounted away from the material as long as a line of sight or mirror "view" can be established. These instruments are sensitive to infrared radiation only at specific wavelengths. Because it is necessary to monitor the temperature all across the width of the sheet, either a number of point sensors must be

TABLE 2.2
Thermoforming related temperatures.

Material	Heat deflection temperature			Thermoforming temperature				Melting temperature		Abbreviated terms ASTM D 1600-92
	at 264 psi °F	at 66 psi °F	no load, material drooping °F	sheet temp. °F	mold temp. °F	plug temp. °F		amorphous polymers T_g °F	crystalline polymers T_m °F	
Polymethyl methacrylate, PMMA										
extruded	200	—	—	275–350	150–170	250–260		185–220	320**	PMMA
cast	205	—	—	290–360	150–170	—		195–220	—	PMMA
Cellulose esters										
cellulose acetate, CA	111–195	120–209	—	260–360	160	—		—	445	CA
cellulose acetate-butyrate, CAB	109–202	130–227	250–300	290–350	175	—		—	285	CAB
cellulose propionate, CAP	111–228	147–250	—	260–360	190	—		—	375	CAP
Polyolefins										
low-density polyethylene, LDPE	—	104–112	—	250–350	150	—		—	230	LDPE
high-density polyethylene, HDPE	—	175–196	220	300–375	190	340		—	275	HDPE
polypropylene, PP	120–140	225–250	285	310–390	190	—		—	345	PP
polymethylpentene, PMP	120–130	180–190	—	500–550	170	—		—	455	PMP
Styrene polymers										
polystyrene, PS	169–202	155–205	225	290–350	150	200		165–220	450	PS
biaxially oriented polystyrene, OPS	—	—	—	350–380	120–150	240		185	—	OPS
high-impact polystyrene, HIPS	170–205	165–200	—	340–360	120–150	200		200–220	—	HIPS
styrene-acrylonitrile copolymer SAN	203–220	220–224	—	430–450	—	—		230–280	—	SAN
acrylonitrile-butadiene-styrene cop., ABS	170–220	170–235	200–250	250–370	160–185	—		190–250	—	ABS

28

TABLE 2.2 (continued).

Material	Heat deflection temperature			Thermoforming temperature			Melting temperature		Abbreviated terms ASTM D 1600-92
	at 264 psi °F	at 66 psi °F	no load, material drooping °F	sheet temp. °F	mold temp. °F	plug temp. °F	amorphous polymers T_g °F	crystalline polymers T_m °F	
Halogen containing polymers									
polyvinyl chloride, rigid, PVC	135–170	140–180	—	250–350	120	180	220	415	PVC
fluorinated ethylene-propylene cop., FEP	—	158	—	450–600	300	—	—	525	FEP
Polyethylene terephthalates									
glycol copolymer, PETG, APET	145	158	—	265–325	130–140	180–190	180	490	PETG
crystallizable, CPET	150	167	—	320–360	250–300	—	180	490	CPET
Modified polyphenylene ether, PPE, PPO									
low glass transition	185–210	230	—	320–380	210	—	210–230	—	PPE
high glass transition	265	280	—	370–410	—	—	245–285	—	PPE
Other engineering polymers									
polyacrylonitrile, PAN	164	172	—	260–360	180	—	203	275	PAN
polyamide 6, PA6	155–185	347–375	—	420–460	200	—	140	420	PA6
polycarbonate, PC	250–270	280–285	—	360–420	250	290	302	445	PC
polysulfone, PSU	340	360	320	450–500	300	—	374	—	PSU
polyetherimide, PEI	390	410	—	—	—	—	420	—	PEI
polyethersulfone, PES	390	410	—	525–650	400	—	445	—	PES
polyether-etherketone, PEEK	388	482	—	750–800	350	—	—	633	PEEK
polyamide-imide, PAI	532	—	—	675–800	450	—	527	—	PAI

*Listed values have been selectively taken from materials suppliers' brochures. Wide variations are possible due to differences in materials and in processing conditions. Refer to Appendix D for temperature conversion chart (page 230).
**Biaxially oriented.

arranged in a row or the installation must be equipped with a sensor having a 90° rotating mirror.

A low-cost method for establishing surface temperatures can be utilized on occasion, especially when setting up or troubleshooting a process. Both temperature-indicating paints and temperature-sensing strips can be applied to the plastic surface. The original matte or white spots will either become clear, become transparent, or, in the case of the strips, show the blackness of the paper when the stated temperature has been exceeded. One such product is supplied by Solder Absorbing Technology Inc. (Springfield, MA) under the trade name CelsiStrip. Other temperature sensors are working on the principle of color change. Coatings are superior to the radiation absorbing paper strips because they are more in intimate contact with the plastic.

In many cases the knowledge of the exact temperature is of less importance as long as one finds an easily recognizable peculiarity correlatable to the material's temperature. Judgments for having obtained the right forming temperature might be made based on the flabbiness of the sheet or the extent to which the sheet sags during the heating process. Some control devices utilize the sagging phenomenon and terminate heating when signaled by a photoelectric cell. This criterion, however, cannot be indiscriminately applied to all plastics, because some materials may already be overheated when they start to sag.

A once established optimum heater output adjustment set by a timer may not repeatedly result in the same sheet temperature. Minor line voltage fluctuations, ambient temperature changes, or sheet gauge fluctuations may significantly alter the actual sheet temperature. Multiple heat-sensing devices inside the heater tunnel are used extensively for control purposes. Even though their indicated temperature does not coincide with the sheet's temperature, the relative temperature changes detected by these devices might still be indicative of the aberrations found in the plastic's temperature.

Where infrared sensors are employed for correct temperature determinations it is important to know that, for practical purposes, the emissivity of any heated plastic is identical to the fraction of total radiation being absorbed by the plastic at the same wavelength. The intensity of this emittance can be measured and easily converted to temperatures. These instruments must be calibrated with nearly identical plastics (chemistry, color, and film thickness). In certain spectral areas interferences may occur due to the absorption of the emitted radiation by water (H_2O) vapor or by carbon dioxide (CO_2) in the air path. In the near-infrared area, readings could be falsified when high-temperature tungsten filament wires are used, due to their reflected, near-infrared radiation being detected by the sensor. The other condition that must be met is that the wavelength of the sensor coincides with a total or high-absorption (zero

Heater controls

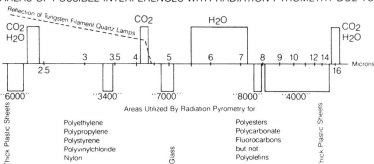

Figure 2.14. Spectral response regions for radiation pyrometry (adapted from MOD-LINE Infrared Thermometers, Ircon, Inc., Skokie, IL 60077).

transmission) region in the infrared spectrum of the plastic in use. When employing copolymers such as acrylonitrile-butadiene-styrene, it is sufficient if one of the main components will absorb in that region.

In the upper part of Figure 2.14 the wavelength regions where interferences may occur are graphically illustrated; in the lower part the wavelength regions of some commercial sensors are shown, and the materials related to them are listed.

Heater controls

To control the temperature of the plastic sheet or film, two avenues can be taken. In simple cases the heaters are powered at a constant energy level, and the desirable sheet temperature is obtained by controlling the heating time of the sheet or the advancement rate in case of automatic roll-fed thermoformers. If acceptable production cycles can be maintained, there should be nothing wrong with such a simple arrangement.

In those less sophisticated thermoformers the heaters are directly powered by the electric line voltage. The temperature of the heater surfaces will rise until oven losses will equal current input. Thermally well-insulated ovens overheat unless plastic sheets are continuously being fed into it. Therefore thermal overload relays must be present in all heaters. Until equilibrium is reached, the operator must individually decide on heating time. Furthermore, small line voltage variations must be corrected by heater time adjustments. Similarly, heater output—but not necessarily energy efficiency—is lowered in time when the electrical resistance of the heater wires or bands increases due to oxidation of the conductor or when heater and reflector surfaces become contaminated. Identical replacement heater elements will have a higher wattage than adjacent older heaters.

In general one can state that, in resistance heating, all electrical energy is converted into heat. The efficiency with which this energy is conveyed to the plastic to be heated is mitigated by the amount of heat conducted or radiated to the machinery and mainly by convection of heated air. Unfortunately, these losses can range from 50 to 90%. The claims of some heater manufacturers showing that their elements provide 96% energy efficiency do not make allowance for such cited losses. Losses occur mainly in the overall equipment and not in the heater elements. The abounding information published in regard to energy efficiency, energy cost, applicability, and cost of heaters for thermoforming must be considered biased or pertain to special circumstances only. The fact that many thermoformer producers offer a choice of heaters confirms that, at present, still no clear winner has emerged.

When thermoforming takes place on automatic machinery and when production scheduling requires a certain output, cycle times must be held constant. In such cases heater output, which can compensate for changes in line voltage, sheet gauge variations or fluctuating heat losses, must be established. Two basic electrical controls have found applications.

In a less sophisticated *open loop control* mode the temperature of the heaters can be adjusted by an operator via percentage timer controls to maintain a constant output. For a *closed loop control* system it is necessary that the power-switching device either obtains a signal from the heat-sensing thermocouple in the heater so that its temperature stays constant or, preferably, that an infrared sensor scans the heated sheet to make sure it remains constant at the desired level without the intervention of an operator. In both cases simple mechanical contactors, mercury switches, solid-state relays, or silicone-controlled rectifiers are being used as switching devices. Applicable controllers would be temperature controllers and programmable controllers.

With the increasing spread of computer technology the possibilities for using infrared sensors for measuring film and sheet temperatures have been expanded. The operator can not only display actual temperatures on a computer screen and compare those values with set data but can also, by means of an operator interface, program temperature profiles or set up the heating process. Apparatus manufacturers are responding to these needs by supplying heater banks with individual heat control and monitoring indicators for each element, thus making it also easy to spot burned-out heating elements.

For example, one thermoformer is keeping process data under control by applying computer-integrated manufacturing (CIM) methods. A programmable controller (PLC 5 from Allen-Bradley Co.) is logging 140 data points each cycle. Most of them relate to temperatures, both for the heaters and the molds. Other data keep track of press movements, indexing strokes and constancy of vacuum, always comparing actual

Figure 2.15. Articulating clamp frame (courtesy of Brown Machine, Beaverton, MI 48612).

values with the machine set upper and lower limits. By this means troubles due to deteriorating seals, plugged vacuum lines, electrical brownouts, and the like can be quickly identified.

Clamping of sheets

For sheet or film heating, as well as for the forming and stripping process, the plastic stock must be restrained firmly between two frames. These frames usually consist of profile irons to which nonslip gripping surfaces are mounted (coarse abrasive cloth, rough rubber pads, weld

Figure 2.16. Infeed view of thermoformer. Both sides: pin chains running on water cooled rails. Center: bottom half of preheater, top half moved to the side.

Figure 2.17. Gripping clips. Mounted on film transport chain of thermoformer (courtesy of Kramer & Grebe GmbH, Biedenkopf-Wallau, Germany).

spatter, etc.). These frames are pressed together by C-clamps, toggle clamps, air cylinder operated jaws (Figure 2.15), and cam or spring-held devices. Sufficient clamping force must be provided to prevent the liberated internal strain, frozen in during the extrusion of the sheet, from pulling the sheet out of the frame during reheating.

Roll-fed automatic thermoformers grip the sheet securely at the sides by means of spring-loaded pin chains running in chain rails (Figure 2.16) or gripping discs mounted on the transport chain (Figure 2.17).

On equipment where only a few pieces a day are formed, it might become necessary to heat the clamping frame so that the sheet material at the flanges will also heat up. The heating of these frames may also be necessary to obtain a better grip of the sheet. In other cases, such as the chain rail of an automatic thermoformer (Figure 2.16), water cooling is provided to keep sheet-gripping pins at tolerable temperatures.

THREE
Thermoforming Molds

DIVERSE AND UNIQUE kinds of molds are used for thermoforming. As a matter of fact, the low cost of molds and the short lead time required for tooling up have led to this forming method being favored over others in many applications.

Generally, only one side of a mold is required, which—depending on the shape of the formed part, the desired appearance, and the process used—may be a male, or positive mold for drape forming or a female, or negative mold for cavity forming. The determination of which one to choose becomes more critical the deeper the part to be formed is. When forming shallow, low-profile parts, the reduction in wall thickness is minimal; therefore, the selection will depend more on appearance. If fine mold details must be duplicated, then the side of the plastic sheet which touches the mold surface should ultimately be the one that becomes visible. Sometimes, a more rounded or smoother appearance is desirable, or the sheet material may already have a pleasant textured surface that could be affected when it touches the mold. In these instances the side that does not touch the mold should eventually be the one that becomes visible in the formed part. It must be realized that a closer dimensional control will be obtainable at the mold surface side.

Reduction in wall thickness: male and female molds

Under all thermoforming conditions in which pieces are shaped from a flat sheet or film, the surface area must become larger and, therefore, the gauge thickness thinner. One of the decisive factors for this thinning is the *draw ratio*. Many times a draw ratio of 3:1 just means that the thickness of an area of a part is just one-third of the original sheet thickness. Unfortunately, three different ways to define that ratio with a comparable numerical value exist, and in each of these cases the values depend also very much on the specific shapes of the part formed.

(1) For irregularly shaped objects the draw ratio is difficult to establish

in a numerical form unless the draw ratio is considered to be the ratio of the maximum cavity depth to the minimum span across the edges of the unformed sheet.

(2) The *areal draw ratio* is expressed by the ratio of the original sheet area within the mounting frame to the surface area of the part after thermoforming. Although these values relate well to the biaxial stretching, they give only an indication of the average part thickness.

(3) The *linear draw ratio* is described by the ratio of the length of a projected line passing through the deepest depression of the thermoformed part to the length of that line on the original sheet. It gives an indication about the highest unidirectional stretching the sheet could have to endure. However, unidirectional stretching is usually restricted to the area close to the frame, because central areas are mainly biaxially stretched.

The difference between unidirectional and biaxial stretching can best be visualized by looking at a long trough. Although both ends must have been formed by biaxial stretching, the central part is only formed unidirectionally. Therefore, differences in wall thickness along the length of the bottom of the trough will be observed. The edges and corners at both ends will be remarkably thinner.

Rendering judgment based just on the numerical values of these draw ratios remains of questionable value, because stretching and thinning is always nonuniform and increasingly progressive with the advancement of the forming process steps. Furthermore, each shape yields another number. Three simple examples should be cited to illustrate how these numbers relate. The values in the table refer to a cubical container formed from a square sheet, a hemispherical bowl, and a cone formed from a circular sheet.

Formed Shape	Common Draw Ratio	Areal Draw Ratio	Linear Draw Ratio
Cube	1	5	3
Hemisphere	0.5	2	1.57
Cone (60°)	0.87	2	2

The thinning of the wall thickness of formed parts having those shapes could result—in severe cases, when forming a sharp cornered cube—in punctures at the four lower corners. When forming a cone, the puncture would occur at the tip of the cone. Due to absence of sharp corners in the case of the hemisphere, one would observe at the apex a reduction to only one-third of the original thickness. The latter finding is shown in Figure 10.3 (page 161). These facts must always be considered when designing a part, because during all vacuum and pressure-forming processes a free forming of rounded shapes, not yet touching any mold surface, will occur throughout the forming process—though at increasingly diminishing areas—up to the point the sheet contacts the mold at the last spot. This stands in contrast to literature citations indicating that those rounded caps are of uniform thickness.

It is important to realize that the reductions in wall thickness are independent from the mechanical properties of the plastic material. Therefore, just changing materials will seldom improve thickness distribution. The search for improvements should thus be centered at reducing depth of draw, avoidance of sharp corners, and mainly by prestretching of the sheet (billow forming and plug insertion).

In general, the polyolefins are most suitable for deep draws, followed by the styrenics, cellulose esters, and polycarbonate. The lowest performers are the foamed or biaxially oriented sheet products and the cast polymethyl methacrylate sheets.

For practical purposes, to determine and compare the draw capabilities of materials, molds that consist of several cylindrical cavities of identical diameter but increasing depth are being used. To obtain practically useful data of materials with only one mold, a female funnel mold with a 60° angle (1.15 diameter to depth ratio) has been recommended by J. G. Williams (Stress Analysis of Polymers, John Wiley & Sons, 1973, pp. 211–220).

To estimate the thinning that may occur, one should determine only the area of the sheet available for forming and divide it by the area of the finally formed part, including trim. For yield determination and for cost considerations the areas that rest in the clamping frame must also be taken into account. Any production piece that yields a wall thickness below specification becomes a reject, and pieces exceeding specified thickness bear higher material costs and longer production cycles (mainly cooling times).

Throughout this chapter the three-dimensional illustrations will show molds or parts with sharp corners for better recognition of the shape. Actual parts and molds should always have generous radii. The example chosen represents a rectangular cavity with side flanges (see Figures 3.1 and 3.2). First, one should consider the ideal wall thickness, assuming the material was uniformly stretched throughout the whole part:

$$\frac{\text{Final material thickness}}{\text{Original material thickness}} = \frac{\text{Available sheet area}}{\text{Area of the total formed part}}$$

$$= \frac{A \times B}{A \times B + E(2C + 2D)} = \frac{3 \times 4}{3 \times 4 + 1(4 + 2)} = \frac{12}{18} = 67\%$$

If the part were made using simple drape forming over a male mold,

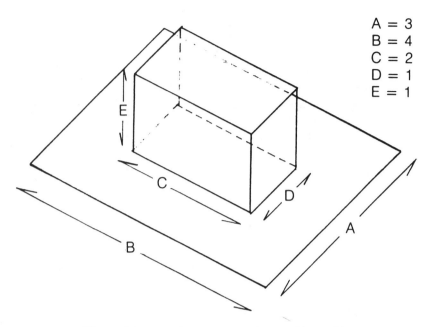

Figure 3.1. Drape forming over male or positive mold.

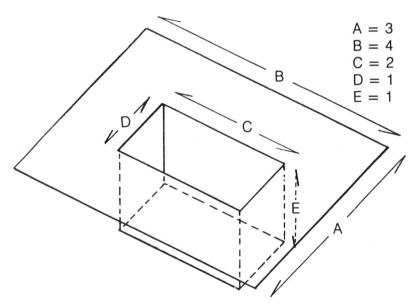

Figure 3.2. Cavity forming in female or negative mold.

the hot sheet would first contact area $C \times D$, which would nearly retain its original thickness. In this case the remaining available sheet area is $A \times B - C \times D$, and the area to be formed out of that piece is found in the denominator of the following equation:

This would be the best available thickness for the rectangular side walls.

$$\frac{\text{Available sheet area}}{\text{Area of part still to be formed}} = \frac{A \times B - C \times D}{A \times B - C \times D + E(2C + 2D)}$$

$$= \frac{10}{16} = 62\%$$

If the same part were formed in a female mold (Figure 3.2), the sheet would first contact area $A \times B - C \times D$, resulting in a flange of heavy wall thickness and leaving only area $C \times D$ for forming the total cavity area $E(2C + 2D) + C \times D$. Therefore, the thinning of the sheet in the cavity would be:

$$\frac{\text{Final material thickness}}{\text{Original material thickness}} = \frac{\text{Available material area}}{\text{Area to be formed out of it}}$$

$$= \frac{C \times D}{C \times D + E(2C + 2D)} = \frac{2}{8} = 25\%$$

In reality none of the vertical or bottom areas will have the calculated thickness uniformly distributed because these walls are not formed at the same time but in sequence, as shown in Figures 3.3 and 3.4. From a line of mold contact the sheet next touches only a narrow part of the mold adjacent to it, and, only after progressive stretching and thinning, continues to form these sides until finally the inside corners fill. These last-formed edges have the thinnest wall thickness. Consequently, calculated wall thicknesses can be considered only as a starting point.

To determine thickness variations, the formed prototype part can be cut into small pieces—or small discs can be punched out of it—and their thickness determined with a micrometer. Other methods include using a translucent or transparent colored sheet, where the differences in color intensity correlate to the thinning of the sheet. The photo in Figure 3.5 illustrates how the thinning of a colored film can easily be spotted. The dark line on one side wall shows a "mark-off" line caused by the premature solidification of the film due to too low mold temperature.

Another very useful approach is to draw parallel lines horizontally and vertically on the sheet prior to forming. Changes in the area of the squares after forming will give information about the material thinning and will also show the extent of draw in any one direction on oblong rectangles

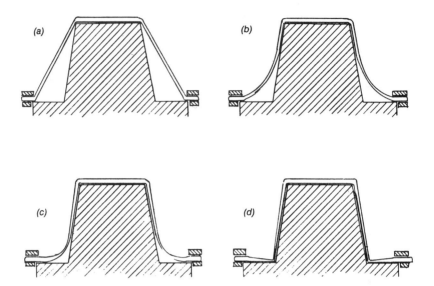

Figure 3.3. Drape vacuum forming over male or positive mold.

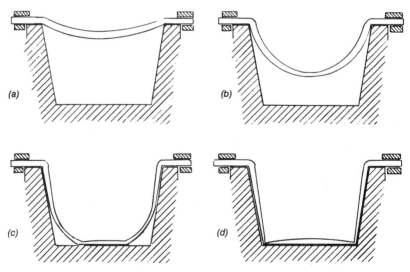

Figure 3.4. Cavity vacuum forming in female or negative mold.

Figure 3.5. Visualization of film thinning, including a "mark-off" line, when using a colored film.

or any other irregular shapes. A variety of thickness measuring instruments, many of those commonly used for paint thickness determination, can also be used for thin parts. These are mainly based on measurements of magnetic or electrical inductive forces and require, therefore, a good contact with a metal piece on the opposite side. Ultrasonic thickness measuring devices that do require access to only one side of the part are now available.

A consideration to be kept in mind when deciding between using a male or female mold is the possibility that webs may form when two adjacent sides of a hot sheet draped around a male mold are closing in over the protruding mold (Figure 3.6). If the hot sheet cannot contract from the A to the B dimension, the excess material will form webs. The more rounded the edges are the less likely that webbing will occur.

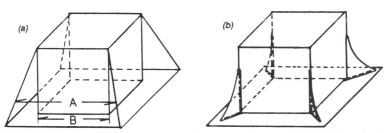

Figure 3.6. Webbing of formed part. (a) Sheet draped over mold. (b) Webs forming when after draping the extended dimension A cannot contract to final dimension B.

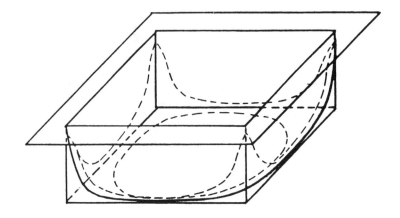

Figure 3.7. Female mold forming. All areas will be under tension in all directions during the forming cycle. Solid line shows rounded contour of not completely formed part. Dashed lines mark contours where sheet misses contact with mold.

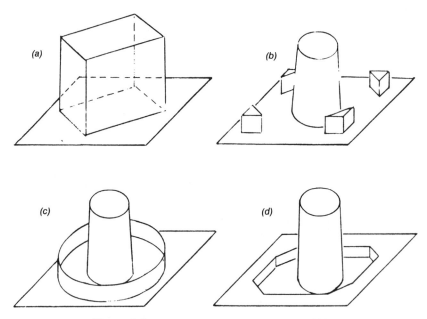

Figure 3.8. Mold arrangements to prevent webbing.

In the case of a female mold the opposite occurs. The plastic sheet will be stretched apart until it contacts all four vertical mold surfaces (Figure 3.7), resulting in extremely thin-walled corners. Again, the rounding of edges will help to retain sheet thickness at tolerable values.

In Figure 3.8, several other means for eliminating webbing are illustrated:

(a) Position square or rectangular parts diagonally on vacuum box. When a snapback box is applied, the same effect can be obtained by equipping it with four 45° gussets in the corners, instead of the 45° rotation of the mold.
(b) Use web catchers or web pullers. These are smaller parts of variable shape that utilize a portion of the excess corner material.
(c) Mount a ring assist (approximately 2 times the diameter of the mold) and one-eighth its height to the vacuum box.
(d) Use a combination of male and female molds. In effect this is equivalent to lowering the mold height by the amount the base is dropped.

Techniques of preventing webbing are not limited to changes in mold design. At lower sheet-forming temperatures the sheet will retain more tenacity and rubberiness, enabling it to more easily retract during the latter stages of forming, which should also be slowed down somewhat so that the material has more time to contract. For large parts, increasing the temperature in localized areas could be beneficial. Using extruded sheets with a higher drawdown ratio (higher orientation) can also reduce the likelihood of webbing. Sheet materials that contain a small amount of cross-links will also reduce webbing.

Multiple male molds must be spaced sufficiently apart to prevent webbing. A distance of 1.75 times the mold height should be adequate. In many instances, to save material, male molds are located closer together. In these cases a rod or strip fastened to the upper clamping frame must be used to separately form each part. One such illustration is shown in Figure 10.6. Automatic roll-fed machines, which have no upper frame, use the identically performing grid-assist male forming technique.

Interchangeable mold cavities are used for many plastic processing methods, but none of them can parallel the benefits achievable with thermoforming. Because thermoforming molds are neither subject to extremely high pressures nor require tight tolerances to prevent penetration of a liquid resin, which could bond the various parts together, adjusting mold depth or adding dividers for a certain overall shape has become standard procedure. The photo in Figure 3.9 pictures five different plate shapes all formed in the same basic plate mold, whose configuration was changed by easily replaceable inserts. Figure 3.10 contains

Figure 3.9. Plates formed in molds with quickly interchangeable bottoms (courtesy of Gabler Maschinenbau GmbH, Luebeck D-23512, Germany).

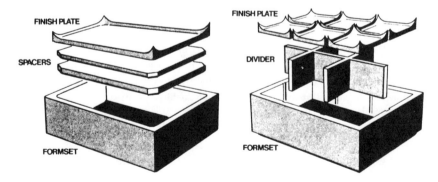

Figure 3.10. Dividable formset and depth adjustment tooling.

Figure 3.11. Quick locks (designed and manufactured by Edward D. Segen & Co., Inc., Milfort, CT 06460).

the drawings for two formsets, which can divide the cavities or adjust the volume of packaging containers by the insertion of various spacers. Faster changeovers are possible if mold cavities with motorized bottom adjustments are selected.

For just-in-time deliveries it is often necessary to quickly replace one mold with another one. Industry is supplying a variety of quick change locks that can be quickly actuated either mechanically or by low-voltage electricity or pneumatically. Figure 3.11 illustrates such components of which the bushings are mounted to the forming table and the locator pins to the mold.

Computer-aided engineering for thermoforming

In the past decades the application of computer programs has greatly facilitated not only the design of injection molds and the selection of materials but also the injection molding process itself. Excellent software exists now that allows designers to forgo costly prototyping, improve mold fill, optimize cooling times, and avoid sink marks, air traps and excessive flashing.

In thermoforming, however, circumstances are more complex. The *finite-element method*, when applied to injection molded parts, allows a modest grid density to be selected and still receive satisfactory information. The finite elements usually consist of triangular shapes to allow better representation of curved edges. In deep thermoforming the most desirable information must be searched in a highly stretched area that occupies only a small speck on the original sheet. Therefore, the analysis must be conducted in several steps so that the system is not overloaded with an unmanageable number of elements nor misses out on important detailed information.

Condensing the model by neglecting the sheet thickness (the third dimension) and by assuming membrane formulations is seen as an acceptable simplification, because bending forces are only encountered at the sheet edges.

The viscoelastic behavior of all the thermoplastics used for thermoforming presents the greatest stumbling block to obtaining an all-encompassing solution. This process is carried out at elevated temperatures (250 to 750°F) where the modulus of elasticity is approximately 10 to 150 psi, reflecting also the applied forming pressure and at forming rates around 1 second. The forming predominately takes place in a biaxial (not necessarily equibiaxial) extension mode. Unfortunately, physical data that could be inserted in such mathematical equations, have not been gathered on those polymers and are difficult to obtain experimentally.

The purely elastic, highly nonlinear behavior of rubbery films can adequately be mastered by existing software due to the power of today's computers in performing repeated iterations. However, in thermoforming of plastics, predominately elastic deformations only take place at low distortions (low draw ratios) and very high rates (fraction of seconds). During deep drawing and thinning of the sheet, deformations are becoming increasingly more viscous rather than elastic in nature, requiring the consideration of a new slate of complicated equations. Probably an intermediate model, incorporating experimental observations with simplified formulations will advance the utilization of computer programs in the future.

In the blow molding process the stretching and accompanying thinning of a heat-plasticized tube has equally great importance. But due to their much higher production volume more research can be undertaken to get a better grip on wall thickness control.

At present there are three software products on the market. Polydynamics, Inc. (Hamilton, Ontario, Canada) is offering T-Formcad 5.0, AC Technology North America, Inc. (Louisville, KY) offers C-MOLD Ther-

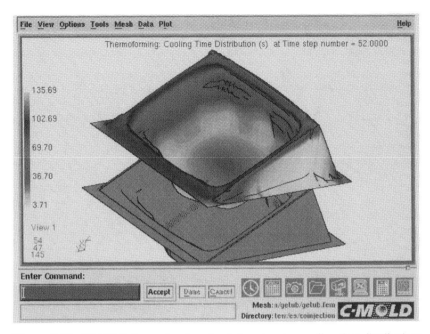

Figure 3.12. C-MOLD thermoforming simulation showing cooling time distributions for a sheet formed over a tub mold. The black-and-white reproduction limits the original's perceptibility (courtesy of AC Technology North America, Inc., Louisville, KY 40223).

moforming (Version 97.7) derived from C-PITA and Accuform (Zlin, Czech Republic) that has T-SIM 2.2. All three programs allow the user to create and rotate a 3-D mold on the computer screen, apply pressure on one side of a heat-plasticized sheet, and observe in steps how it deforms until all parts of the sheet eventually contact the whole mold surface. The thinning of the sheet is quantitatively marked by a color scale from red for the heaviest sections to violet for the thinnest parts. The required cooling time can similarly be indicated, as shown in Figure 3.12. Due to the lack of color reproduction, further explanations are necessary. Only the two small light gray triangles at the left and lower corners represent the areas of thickest cross section (nearly unstretched), requiring approximately 130 seconds cooling time. The areas of the lower outside wall (shown on the right side) are the thinnest (approximately 1/32") and, therefore, cool within 4 seconds.

These products are based on and limited to the hyperelastic material model as described by Mooney-Rivlin and Ogden. The areas of excessive thinning can be traced back to the original flat sheet, making it possible to optimize wall thickness distribution by profiling sheet temperatures. This is performed by screening or by varying area heater output. The time the expanding sheet needs for touching the mold surface can also be determined. Areas where webbing is likely to occur with male molds are becoming noticeable by kinks in the wire mesh.

Part shrinkage and dimensional tolerances

Part shrinkage and dimensional tolerances are quite different for parts formed on male molds and those formed on female molds. On a male mold (Figure 3.3) shrinkage is prevented as long as the part is allowed to cool on the mold. If cooling to ambient temperature could take place on the mold, elastic shrinkage in the part would be minimal. That is, part dimensions would correspond very closely to mold dimensions.

The fact is, however, that on male molds the parts must be removed while still warm, because, otherwise, they would strangle the mold and become difficult to extract. In addition, the time required for cooling to ambient temperatures is unacceptably long. Removed while still warm, the formed part will unavoidably be smaller than the size of the male mold on which it is formed. This is referred to as thermal shrinkage, which is proportional to the difference between the ambient temperature and the temperature of the part as it is removed from the mold times the coefficient of thermal expansion. Thus, to maintain a specified part size, it is necessary that the mold be slightly oversized and that the part be taken from the mold according to a predetermined time schedule. The sheet-forming temperature and the rate of forming are practically of no consequence.

With the comparable female mold (see Figure 3.4), the formed part begins to shrink as soon as the material temperature falls below the setting temperature, which could be close to the heat distortion temperature or the onset of crystallization. Shrinkage is not restricted by the mold; therefore, the duration of the cooling cycle in the mold is of no consequence. The shrinkage will be higher than with the male mold due to less interference during cooling. To maintain close tolerances repeatedly, the mold temperature must be kept constant and the magnitude of the vacuum (or pressure) force maintained consistently at a higher level and for a controlled duration. Because wall thickness is also influenced by the forming conditions, it is necessary to maintain the process within much stricter forming parameters. Variation in sheet thickness will additionally affect tolerances of inside dimensions.

The maintenance of tolerances over long thermoforming production cycles, that is, the *consistency in part dimensions* is much more difficult to achieve than with other plastics molding processes. The production contract should specify as few limits as possible and only where dimensions are really critical. Otherwise, extra people will have to be assigned for making measurements.

The designer must put some ledges, protrusions, or dimples into the mold, not only for providing reference points for taking measurements but also to keep critical regions better constrained during cooling. These markings are especially helpful when cutouts or holes are incorporated into the part. A positive support is also required if accurate trim dimensions must be upheld.

Because maintainable tolerances can vary so widely with the kind of equipment and material, no specific numerical data are given here. Past experience should be the best criterion. As a guideline, mold shrinkage should be assumed to be approximately 0.005 in./in. for male molds and nearly double that for female molds. A lower mold shrinkage is found in the cellulosics and rigid polyvinyl chloride (0.0035 in./in.); high mold shrinkage materials are the acrylics, polycarbonate, thermoplastic polyesters, and oriented polystyrene (0.008 in./in.). Materials with excessively high mold shrinkages include all the crystalline materials, such as high-density polyethylene, polypropylene, and the fluorocarbon polymers (0.025 in./in.). However, these numbers must be used cautiously because the following conditions may alter them significantly:

(1) Mold temperature: A 15°F temperature difference can change shrinkage by up to 0.001 in./in.
(2) Size and shape factors: These refer to the degree to which the part is restrained in the mold and to the effect a greater wall thickness has on the temperature profile.
(3) End-use temperature: Due to normal expansion or contraction proportional to the coefficient of thermal expansion, part dimensions continually vary with changes in ambient conditions.
(4) Extreme use conditions: Shrinkage will reach peak values after first exposure to the highest use temperature. This is especially important to consider when heat sterilization is applied.
(5) Sheet orientation of extruded sheets: Drawdown during extrusion might cause a shrinkage differential of 0.002 in./in., with the shrinkage in the machine direction being higher than in the transverse direction.

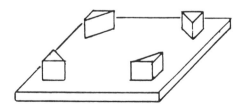

Figure 3.13. Cooling fixture for rectangular part.

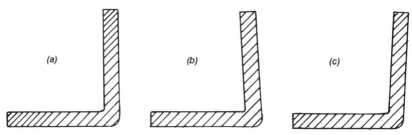

Figure 3.14. Angle pieces 90°. (a) Shape of mold used. (b) Injection molded part extracted at too high temperature. (c) Thermoformed part extracted at too high temperature.

Where part dimension tolerances are less than 0.005 in./in., it is advisable to build somewhat oversized molds in the case of male molds and undersized in the case of female molds. This is so that subsequent corrections can be made without the need of adding material to the mold surfaces.

Sometimes it may be necessary to restrain shrinkage and prevent distortions by placing larger formed parts over cooling fixtures until the part is completely cooled to ambient temperature. A cooling fixture for a part similar to Figure 3.6 could be as simple as shown in Figure 3.13. If too many dimensions become critical, one might have to choose another forming method, such as pressure forming or injection molding.

The foregoing precautions do not guarantee that the formed part will not later become warped or dimensionally changed, especially with crystalline materials. Also, parts that have been forced into shape at relatively low forming temperatures may tend to revert from the imposed shape in time, due to the plastic memory of the material. This is more manifested for parts that have only been subject to shallow deformations. One should bear in mind that these changes are generally the opposite of those encountered in injection molded parts. This is illustrated by the (somewhat amplified) 90° angled parts in Figure 3.14. But such a picture would appear if the formed or molded part would be cut into thin slices.

Because of these variations it is imperative to test formed parts prior to mass production. Shrinkage and warpage should be monitored on samples of the formed parts, e.g., 1 hour, 1 day, and 1 week after forming. Testing should also be done after exposing one set of parts to the highest use temperature and another to 100% relative humidity (plastic bag with some water inside) for at least 24 hours.

Warpage

Many times warpage can occur unexpectedly or sporadically. Non-

Warpage 51

symmetric and unbalanced molds are more likely to cause warpage. Any imbalance in any of the following conditions should become suspect: irregular changes in extruded sheet wall thickness due to irregularity in transverse or longitudinal gauge, uneven heater output, unequal cooling of mold, or uncontrolled air draft. Slight shrinkage and thus movement can occur in the parts for 24 hours after forming. Perfectly formed parts can arrive at the customer's locale warped if either packaged too soon or nested improperly so that strain becomes introduced in the parts. In cases in which undue pressure may lead to blocking of stacked parts along the draft sides it might become advantageous to incorporate several denesting lugs into the mold to prevent crushing of stacked parts during transport. On the left side of Figure 3.15 and magnified in the foreground the eight lugs incorporated into a packaging shell are shown. This sample was pigmented to show better details. At the right side the uniform spacing of three stacked clear parts are shown. In the background the folded packaging for a ball-shaped merchandise is shown. The positions of the lugs are alternated on sequential parts.

Another possibility for preventing formed parts from becoming jammed or distorted when stacked and during transport is illustrated in

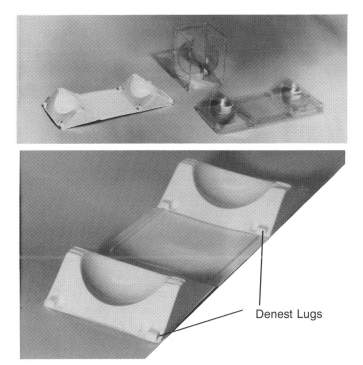

Figure 3.15. Denesting lugs incorporated into thin packaging to prevent crushing or distortions during shipping and handling.

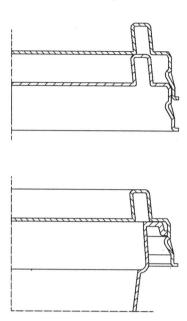

Figure 3.16. Undercut forming can prevent stacking problems (courtesy of Marbach Tool & Equipment, Inc., Elkhart, IN 46516).

Figure 3.16. The ridge around the lid does not only maintain the distance between stacked lids but also ensures good fit and tightness despite minor dimensional aberrations between container and lid.

Draft in the mold

Vertical surfaces should have a 3 to 6° draft angle on male molds for easy removal. If one vertical surface must be exactly at a right angle, the formed part can still be easily removed from the mold when the edges on the sides are either generously sloped or have a copious radius. Due to the high shrinkage of plastics during crystallization, highly crystalline plastics require more draft in the molds. Smooth female molds do not necessarily require a draft. Undercuts, if required for closures or snap-fit assemblies, should be relatively small in area and must be avoided on outside corners. The forming table should be equipped for actuating stripper plates in those areas. For large undercuts, collapsible male molds, two-piece female molds, laterally movable sections, spring-loaded flippers, or unscrewing inserts should be provided for convenient part removal.

Radii at edges and corners should be as generously laid out as possible. They not only ease the extraction of the part from the mold but also improve the flatness of adjacent areas. The appearance of chill marks there can be minimized (see Figure 3.5). Sharp corners can lead to web formation on tall male molds and also carry the danger of brittle failure of the part. Rounded edges improve stiffness, reduce molded-in stresses, and are more likely to prevent warpage.

Surface appearance

The surface details obtainable by injection and compression molding or by casting cannot be produced by conventional vacuum forming. (Pressure forming is described on page 65). This is especially true for gloss. Even highly polished molds, which are never recommended, would not improve gloss on formed parts. Conversely, high-gloss sheet materials may sometimes lose their gloss during thermoforming. In addition, glossy pigmented surfaces are undesirable because they emphasize chill marks, undulations, and changes in wall thickness. The appearance of large flat surfaces can be improved by gently dishing them in a convex shape or subdividing them into smaller areas having slightly altered orientations.

Recognizable details, such as markings, should be at least 3 times the thickness of the thinned out material. Scratches and minor indented mold imperfections will not be transferred to the part. However, dirt and dust, which continue to adhere to the sheet material or the mold, will cause unsightly blemishes. Thus, painstaking cleanliness in the work area is compulsory.

An overly smooth mold surface may pose difficulties in obtaining flat surface contours. The plastic will first touch the surface only in certain areas, and, as a result, air pockets may block off further forming there. These blemishes become magnified at high sheet-cooling rates, especially with crystalline polymers, such as high-density polyethylene, where additional heat of fusion must be conducted away. Blasting the molds with glass beads or coarser abrasives can eliminate the air pocketing because it permits air movement between the mold and the area of the part already formed. Scratching the mold surface with coarse emery cloth is also effective in creating air channels. These mold surface imperfections also help to break the vacuum when, after forming the parts, they are stripped from the mold.

Small lettering or numbers can easily be incorporated into the mold as long as every recess area has its own vacuum access as shown in Figure

Figure 3.17. Mold details for raised characters. Small circles show air vent holes.

3.17. Vacuum hole sizes smaller than the local material thickness will not become visible.

Mold materials

In contrast to most plastic molding processes, relatively low temperatures and pressures are normally employed in thermoforming. For this reason prototype tooling becomes often sufficient to produce the few parts that are needed. Depending on many variables, the material for making such molds can be selected from wood, plaster, casting resins, glass fiber–reinforced plastics, white metal alloys, or reinforced, sprayed, or galvanic deposited metal shells.

Wooden molds, if made of well-dried hardwood, are often used for low production items. The low thermal conductivity of wood makes it ideal, because the heated sheet will not be chilled at the first contact. However, for fast repetitive forming, the heat rejection of hardwood molds is insufficient. Service life and part quality can be improved by high-temperature epoxy varnish coatings.

Inexpensive molds can also be prepared by a casting process using one of the large variety of water miscible plasters. The vent holes can be incorporated by distributing at the desired locations mold release–coated wires, partly inserted in fine tubing. With these materials mold shrinkage could be appreciable. They can be strengthened by incorporation of chopped glass fibers, wire mesh, or steel rods. They must be thoroughly dried and sealed to prevent surface irregularities caused by steam generation when contacted by the heated sheet.

More temperature-resistant molds for rapid forming cycles can be built up of phenolic boards or laminates. They are less affected by heat or humidity.

Mineral or powdered metal–filled epoxy and polyurethane molds are extremely durable and very versatile. With little effort a number of epoxy molds can be prepared from just one wooden pattern. Initially, as many cellulose propionate shells are formed as final molds are needed for the production run. With little preparation these shells can be made to serve as molds for a room temperature curing epoxy-casting compound. These castings can be provided with the necessary air passages and cooling tubes so that they become useable as mold blanks for a lasting multicavity form. The thermal properties of epoxies are well suited for long production runs at slower cycles. If copper cooling coils are embedded in them, their heat removal capability will improve only somewhat, but it still will not suffice for fast mass production runs.

For large molds extending over several square feet, plastic molds are prepared by a hand lay-up or spray-up process using glass fibers as reinforcement and epoxy or unsaturated polyester resins as binder. At first, a smooth coating (often called gelcoat) is applied to the polyvinyl alcohol-coated pattern, followed by a thin layer of a formable fine glass fiber veil. Depending on strength requirements, chopped glass mat or glass fabric layers are bonded and cured to build up the surface layer. A great variety of fillets, crates, or trusses can be bonded with a resin paste to the shell to complete the construction of a sturdy mold. If unsaturated polyesters have been used, the mold surface temperature should not exceed 150°F; epoxy molds can withstand up to 300°F.

The manufacturing time for metal molds was drastically reduced when metal spray processes or galvanic metal deposition on suitable patterns were introduced. One possibility could be to start with the deposition of a thin stainless steel surface followed by a heavy, structurally strong zinc backing; another could be to start with electronickel deposition followed with a thicker layer of copper. In all cases strength has to be built up, preferably with other metals to prevent thermal expansion differentials, which could destroy such molds due to delaminations, especially when used for pressure forming.

Aluminum molds are best suited for the longest production runs. They are either prepared by a casting process or the buildup and machining of stock aluminum shapes. The high thermal conductivity of aluminum makes it necessary to preheat the molds at the start either by circulating hot water through the cooling channels or by exposing the molds to radiant heat. For rapid forming rates it will be necessary to cool these molds. This should be done with thermostatic controls to ensure that mold surface temperature fluctuations are minimal and the mold is not overcooled. Adequate temperature constancy of the molds can only be obtained by restricting the temperature difference of the chiller to a few degrees and increasing flow rate so that coolant turbulence is ensured. Hard metallic coatings or the application of polytetrafluoroethylene can improve the wear properties of aluminum molds, and the latter may also improve draping of the sheet during forming. This can help to reduce sheet thickness variations. If careful use of molds cannot be ascertained, steel molds should be considered.

Porous metal molds have the advantage that no drilling of vent holes is required. Their disadvantages are: prolonged cooling times, increased fragility, and the possibility of reduced porosity over time.

Mold-cooling provisions

In many cases the means for cooling the formed part are incorporated into the mold. Cooling channels, drilled cooling paths, or incorporated cooling pipes must be located properly and must be capable of carrying a sufficient volume of coolant. More details are described in Chapter 5 and shown later in Figure 3.20.

A balance must be established between rapidly cooling the formed part and not excessively chilling the sheet yet to be formed. This discrepancy can lead to imperfections. One should watch for the appearance of waviness on otherwise smooth surfaces and for abrupt changes in wall thickness caused by the inability of the overly chilled sheet to slip over edges or even over flat surfaces.

A number of expedients can help offset this cooling difficulty. Critical mold areas can be outfitted with plastic inserts of nylon or polytetrafluoroethylene. Sometimes a plastic coating applied to the desired areas may be sufficient to reduce thermal conductivity. In some situations the problem can be solved by keeping the sheet of plastic aloft at the region of first contact. This can be accomplished by blowing some air through holes in the center of the mold, until an additional segment of the part has been formed.

Air passage holes

A large quantity of air must be rapidly removed from the space between the mold and the sheet. If a single opening were used, its diameter would have to be 1/2" or 1", depending on the volume of the formed part. A single opening would not work, because as soon as the plastic sheet were drawn over it, the forming of other areas would cease. Therefore air passage holes must be distributed over the entire surface. They should be concentrated at inside edges and corners and spaced linearly at a distance of 1/2" to 1 1/2". The size of the air holes should be small enough not to be replicated on the formed part. Whether this happens is partly contingent on the thickness of the material. As a starting point, air passage holes should have a diameter equivalent to the thickness of the part to be formed, not to the thickness of the original sheet. For very thin and very thick parts this rule does not apply. Depending on the shape of the mold, either backdrilling (as shown in Figure 4.3) with larger holes close to the mold back surface, or wide channels, can be arranged in the back. Thin air passages are then drilled from the mold surface to meet them. Large numbers of vent holes may be incorporated in cast molds by embedding grease-coated fine wires, which are later pulled, or by using a porous casting compound.

The sheet will be draped naturally over outside corners of the mold. No holes need to be located in these areas, because this would only reduce slippage over corners during forming. On the other hand, holes should be packed densely on inside corners and along concave radii. A narrow groove along inside corners is very effective and, depending on the way a mold is built—shimming of stacked mold parts at the plane of inside corners—can facilitate air removal and save the drilling of many small holes as illustrated in Figure 3.18. If lettering or detailed texturing is to show up properly, every small dimple must have at least one vent hole or must be connected by a channel or groove (Figure 3.17). A vapor-honed or surface-scratched mold surface enhances air movement and prevents the formation of unsightly surface air entrapments, which are more likely to occur with thin and limp plastics.

On occasion, one might take advantage of trapped air in parts of a mold where one would like to have excellent optical properties and transparency. The recessed top of the mold shown in Figure 3.19 has no vacuum holes. Air will be evacuated only around the bottom edge. The only disadvantage of having a window untouched by the mold surface is that one must tolerate longer cooling times.

The rate of forming will depend on the total area of all air vents. To quickly remove the bulk of the air on voluminous parts, it may be advantageous to locate a few larger holes (1/8" or 1/4" in diameter)

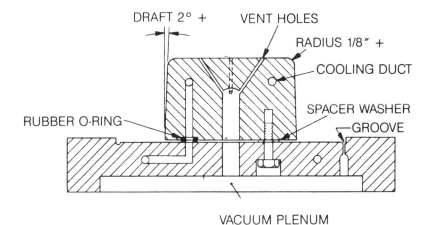

Figure 3.18. Drape mold.

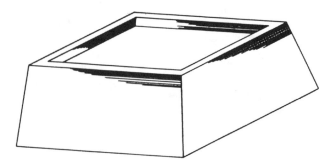

Figure 3.19. Mold with recessed top having no air vents will result in blisters of excellent transparency (courtesy of Eastman Chemical Company, Kingsport, TN 37662).

outside the periphery of the part. These areas will eventually end up as trim. To reduce evacuation time and save on surge tank capacity, the dead air volume inside large hollow male molds should be decreased by filling this space with plastic balls or closed cell foam pieces (polystyrene).

Another important function the mold sometimes shares with the sheet-clamping frame is to secure and stabilize the position of the heated sheet and to sustain a good airtight seal between sheet and mold. For high-density polyethylene sheet, a material difficult to vacuum form, such an addition has proven to be of great advantage. Figure 3.20 shows a mold having a moat (or sometimes a dam) consisting of a shallow groove (at least 1/4" wide and 1/16" deep) machined all around the part, precisely at or outside the trim line. Figure 3.21 illustrates the beneficial effect of such a moat for forming a rectangular pan out of polyethylene.

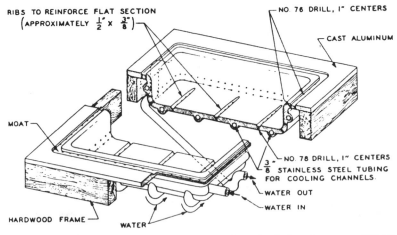

Figure 3.20. Female suitcase mold for high density polyethylene (courtesy of Phillips Chemical Company, Bartlesville, OK 84004).

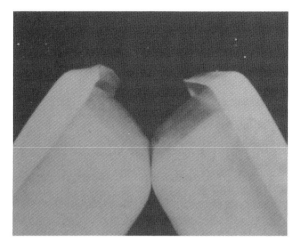

Figure 3.21. Molded rectangular pans. Left side without and right side with moat around the mold (courtesy of Phillips Chemical Company, Bartlesville, OK 74004).

Increasing stiffness

If possible, completely flat surfaces should be avoided, because the slightest distortions will show up and the part will lose its quality appearance. Rounding and arching of surfaces or incorporating high reliefs and ridges increase the stiffness of the formed part and allow the use of thinner sheet materials. Shallowly waffled surfaces will have a similar effect.

Mold plugs

As indicated in the first pages of this chapter, considerable thinning of the sheet may occur when deep drawn parts are formed. In many of the processes described in Chapter Ten, plugs are employed to initiate the forming by prestretching the sheet so that it can thereafter be more evenly distributed in the mold.

If the forming equipment is fitted with a second movable platen, plug assists will remarkably improve the overall thickness distribution in the formed part. Their purpose is to stretch the hot sheet to the rough outlines of the part to be formed. Because final forming takes place after this preforming, the plug must not remove any heat from the sheet and also should not, if possible, mar or scratch the sheet surface.

Plugs can be constructed from the same materials as the molds. Appearance requirements dictate the selection of and expenditure for plugs. They are listed in sequence of increasing "mark-off." The three different categories are:

(1) Metallic plugs, usually made of aluminum, must have a smooth surface and must be kept within a narrow temperature range, usually only 10°F lower than the sheet material temperature. If the temperature is too high, the sheet might stick to the plug.
(2) Wooden, plastic, or metallic plugs built on the principle of good thermal insulation obviate the need for temperature control. Their surface could consist of soft wood, syntactic foam plastic, or any material having a layer of cotton flannel cloth, velvet, or felt draped over the surface. A polytetrafluoroethylene-clad plug will eliminate sticking of the heated sheet.
(3) Skeleton plugs consist only of welded rods, resembling the contour of a solid plug. Corners and edges must be smooth. A heavy plastic coating will improve performance.

The dimensions of a plug in relation to the dimensions of a part will have a great influence on the part's material thickness distribution. It is fortuitous that just by changing the penetration depth of the same plug (usually 75% of the part's depth) the wall thickness ratio between part bottom and part sides can be regulated. Therefore, the equipment should have the capability for adjustment of plug depth penetration and speed of insertion. The latter is especially important in billow forming when the material must roll smoothly into the mold.

The design of a plug depends on the stretch behavior of the plastic. When forming ethylene vinyl alcohol copolymer film laminates, a regularly sized plug is used for the higher temperature melt phase forming process. However, when higher material orientation is wanted, the laminates are formed at lower temperatures (320°F in case of a polypropylene laminate) according to the solid-phase pressure-forming process. In this case most of the stretching is accomplished with the plug, the size of which should be nearly as large as the cavity is. This will prevent cracking of the barrier layer when the initially only axially stretched container becomes oriented circumferentially during the succeeding air pressure forming (at approximately 400 psi).

In some cases the plug may incorporate a crown or ridge to make more material available for a sizable dimple or groove in the mold. Skeleton plugs can be made expandable or capable to perform articulated motions. They can stretch the bottom part of the sheet after insertion, resulting in thin bottoms surrounded by relatively heavy bottom edges. When forming long, narrow troughs, an ordinary plug will still result in thin end regions, which can be overcome only by having the plug actively stretching the material from the center zone toward both ends.

FOUR
Vacuum, Air Pressure, and Mechanical Forces

ALTHOUGH MECHANICAL FORCES are of increasing interest, air pressure is still widely used to accomplish the forming task. It is a gentle force, leaves no marks on the surface, acts in all directions with equal force, and is readily available. The magnitude of this force is established by the difference in the air pressure acting on each side of the sheet. With a vacuum on one side, the maximum pressure available is the atmospheric pressure of 14.7 psia. For most forming processes only a fraction of this pressure is attained. If higher forces are required or if it becomes impossible to establish a good seal at the low-pressure side of the sheet, pressurized air can be used. Other advantages of using pressurized air are described later.

Measuring vacuum and pressure forces

There are two basic ways of expressing and measuring vacuum. One can use gauges that measure absolute pressures. The readings obtained with these more expensive instruments will not change with changes of altitude or weather. In the other case, atmospheric pressure, wherever it may be, is chosen as reference point. The pressure difference or vacuum is expressed in pounds per square inch gauge pressure (vacuum) or in inches of mercury. Both can be accurately converted using the following factors:

$$1 \text{ in. Hg} = 0.4912 \text{ psig or}$$

$$2.036 \text{ in. Hg} = 1 \text{ psig}$$

The force exerted onto every square inch of the sheet to be formed is identical to the psig readings or about half of the value if the gauge reads in. Hg. Therefore, these values should be used for establishing working parameters when setting up operating conditions. Other readings, such as psia (absolute), atmospheres, Torr, etc., should not be used for process controls and logging because they change with location and weather.

Therefore, working parameters should not be selected too close to the vacuum capabilities of the equipment, because these conditions may not be obtainable under adverse low-pressure weather conditions, making it necessary to readjust cycle time.

For any vacuum-formed part one must consider the volume of air that must be removed to retain the actual force sufficient to form the last dimple in the part. Both requirements are counteracting each other. Voluminous parts will require the evacuation of a large amount of air. This may drain the leftover power of the vacuum, resulting in improper figuration of the last detail. On the other hand, shallow parts will be formed quickly, leaving adequate force in reserve. In the first case some power can be saved by filling large voids in male molds with closed cell foam.

Vacuum sources

Many kinds of vacuum pumps can be used. Reciprocating piston, diaphragm, sliding vane rotary, eccentric rotor pumps, etc., all establish a good vacuum but are less capable of evacuating a large volume of air quickly. For this reason they are connected to an air reservoir (surge tank) that serves as a vacuum accumulator. On the other hand, air blowers can rapidly remove large quantities of air, although somewhat limited in the force they can eventually attain due to their confined vacuum. Compressed air-operated air ejectors are ideal in both respects and are very low in cost. They are not found in commercially built equipment. They require a powerful source for compressed air and should be equipped with silencers and a sound-attenuating enclosure.

Vacuum accumulators or surge tanks

With the exception of some of the rarely used vacuum sources, the vacuum for forming the parts will be supplied from vacuum surge tanks that can rapidly evacuate the space around a mold. For the forming of small parts in a roll-fed automatic thermoformer there will be no problem for finding space for a surge tank having a volume that is sixfold the volume of the space to be evacuated. This would correspond to a 15-gal surge tank when forming twenty 16-oz cups under a net forming pressure of 12.5 psig (approximately 25 in. Hg) vacuum. These parameters are, therefore, generally cited in the literature. However, to apply the same conditions for forming a spa, the surge tank would need to have a capacity of 2700 gal. To get the job done with a volume of the surge tank being reduced to 2.5 times the volume of the space around the mold, the vacuum box, and the vacuum lines, a working pressure force of only 10 psig (approximately 21 in. mercury vacuum) would be attainable. Most thermoformed parts will show satisfactory replications of all details at such a pressure setting. Doubling the volume of the surge tank will—all other conditions remaining the same—increase the forming pressure by only 15% to 11.5 psig. As stated, the theoretical upper limit for vacuum forming is only 14.7 psi. To obtain another order of magnitude in detail replication, the forming pressure must be increased fourfold to a level commonly used in pressure thermoforming.

In most cases swift delivery of this force is of greatest importance. This can only be accomplished by locating the surge tank close to the mold and keeping the friction losses in those connecting lines as low as possible. This is achieved by:

(1) Large inner diameters of the lines
(2) Gently curved lines only, with no right angles
(3) No changes or restrictions in cross section

Many pieces of equipment on the market transgress these rules. There is a big difference in conveying a given volume of compressed air through a line versus the same volume of vacuum. A 1-in. line is the minimum for evacuating 1 cu ft of air space. For large parts 2- and 3-in. lines should be considered. To obtain a lasting vacuum connection between surge tank and the moving mold, a flexible, steel wire–reinforced vacuum hose with a comparable interior diameter must be used. Where a slow draw is desirable, an in-line manual ball valve could be considered. Figure 4.1 depicts on the left the arrangement of a vacuum accumulator and on the right (magnified) a vacuum box mounted on a movable vacuum or pressure table platen.

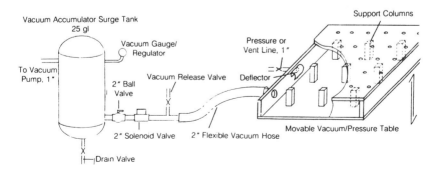

Figure 4.1. Arrangement of vacuum accumulator and a platen mounted vacuum box.

If the forming process includes a pressure-induced billow or prestretch forming step, the surge tank of larger capacity should be selected. Otherwise, the pressurized air must first be vented before the application of vacuum to prevent excessive loss of vacuum from the surge tank.

It is evident that the positioning of the heated sheet in reference to the mold must ensure an airtight seal. This may sometimes be difficult to attain on very tall male, drape formings.

Application of vacuum forces

In applying a vacuum, many variables are encountered. In general, vacuum pumps operate constantly to maintain vacuum in the surge tank. The gauge readings fluctuate with each cycle. On the other hand, blowers and air ejectors are turned on only for the time vacuum is required. They quickly establish a constant pressure differential. In either case the vacuum on the formed part must be maintained until it is sufficiently cooled to resist the material's inner forces, which tend to revert the part to the original shape and may cause warpage.

The faster the vacuum can be applied, the better the appearance of the part. On occasion, a slower forming rate may be called for when large pieces made up of heavy sections are produced. The slower cooling rate in these cases will allow a longer forming time, enabling the material to flow in a viscous mode. When using tall male molds and where webbing becomes a problem, slowing down the forming step and reducing forming temperature may give the plastic more time to contract in the transverse direction, thus eliminating web formation (see Figure 3.6 on page 41).

In many cases the vacuum supply line leading to the forming table can also be used to conduct compressed air to the mold box. After the main vacuum valve between the surge tank and vacuum box is closed, a brief blast of compressed air can detach the part from the mold by the reversed flow of air through the multitude of air passages in the mold. However, better results are obtained—especially with tall male molds—when air pressure is supplied via a separate system, with air exiting only centrally at the top of the formed part.

If for the thermoforming process a billow prestretching step is included, a supply of low-pressure air (less than 5 psi) must be guided baffled into the cavity so as not to impinge directly upon the heated sheet. If the vacuum resources are tight, one should consider first quickly bleeding off the pressure before applying the vacuum for the forming process.

Pressure forming

In applications in which the vacuum force is replaced by an air pressure force from the opposite side, one must consider that it is more complicated to obtain a satisfactory seal for pressurized air. The forming force can readily be doubled or even applied tenfold if compressed air at 100 psi is available. However, ordinary vacuum molds and forms may not be able to withstand such pressures or may deflect excessively. For pressure forming, all precautions applicable to pressure vessels must be obeyed. A medium-sized mold might require a clamping force of several tons, which naturally cannot be provided by C-clamps or toggle clamps.

Pressure-forming equipment must be sturdier than vacuum-forming machinery. Figure 4.2 shows the close-up of a rugged four-point double-toggle linkage used on this type of equipment. The entire machine is shown in Figure 7.10 (page 87).

As in vacuum forming, a tank of comparable size should be available for the compressed air supply. Pipe connections are not critical because pressure drop will be negligible. (If a ducting system has a pressure drop of 5 psi, the pressure loss in a vacuum system of 10 psi is 50%; in a pressure system of 100 psi it is only 5%.) Close to the mold a pressure-reducing valve and a pressure gauge should be installed. It is important to have baffles at the entrance to the mold, so that the cold air will never be directly blown on the heated sheet. Sometimes, the air is guided over heaters, which will help when blowing large billows that must stay hot until the sheet is eventually formed onto the mold.

Figure 4.2. Four-point double toggle linkage of pressure-forming press.

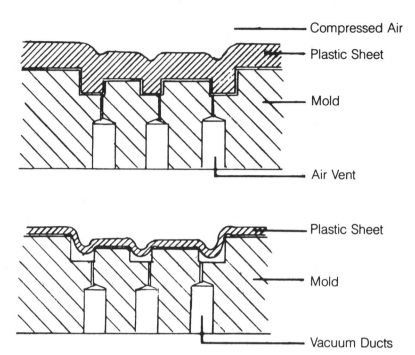

Figure 4.3. Enlarged cross-sectional view of a reproduction detail obtained by pressure (top) and vacuum forming (bottom).

Pressure forming has become popular mainly for small parts. Its advantages are that dimensional tolerances are improved and that forming speed can be increased considerably. It is capable of duplicating fine surface details, such as the lettering or picture images illustrated in Figure 4.3. From the outside it can be difficult to distinguish pressure-formed parts from injection molded parts. Because every mold detail becomes visible in the part, the forms must have qualities close to injection molds. This refers also to venting. All commonly drilled vent holes will show up.

Mechanical forming

As indicated previously, thermoforming is not limited to pneumatic techniques; various mechanical forces may be applied instead.

The simplest application of mechanical forces is utilized in two-dimensional forming. In this case the heated sheet is laid over a curved mold that usually has a soft thermally insulating surface. Gravity will normally suffice to bend the sheet and duplicate the shape of the mold, and no stretching, which would change wall thickness, occurs.

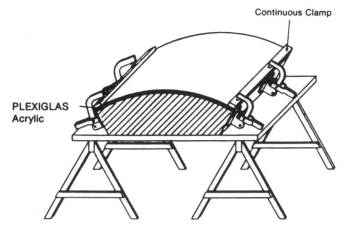

Figure 4.4. Male drape form (courtesy of Rohm and Haas Co., Philadelphia, PA 19105).

The drape form drawn in Figure 4.4 shows that the sheet edges are lightly clamped to keep them straight during slow cooling. When only a narrow part of the sheet must be bent, it is not necessary to heat the whole sheet. Strip heaters, as shown in Figure 4.5, consisting usually of electrical resistance wire coils protected and supported by metal or glass tubes, confine the heating of the plastic to a straight line. The sheets are then clamped between rails and bent at a somewhat sharper angle to compensate for the expected amount of spring-back. Because only a small part of the sheet is heated, stresses will develop during cooling. To keep stresses at a minimum, the formed part should be cooled slowly and uniformly. Excessive stresses may cause warping, cracking, and crazing of the part if it is subjected to severe environmental conditions. For polyvinyl chloride sheets, the previously mentioned dielectric heaters could serve well.

Pairs of matching die halves are used for the most complicated forms. For high-production items the molds are closed and held together by

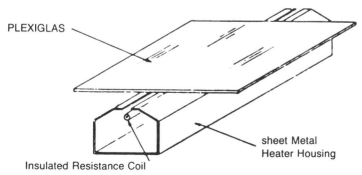

Figure 4.5. Strip heater (courtesy of Rohm and Haas Co., Philadelphia, PA 19105).

pneumatic or hydraulic cylinder pressure utilizing the great force amplification of toggle linkages. In other cases, motion is provided by servomotor-driven eccentric linkages. If both mold halves are temperature controlled, reductions in cooling time, compared with vacuum forming, can be significant. Often, heating time can be curtailed, because higher forming forces are easily obtained. The forming of polystyrene foam packaging represents an important application of this method.

Some of these processes, including slip forming, drawing, or stamping, may be performed at considerably lower temperatures than customary for thermoforming. More details on them are found in Chapter Twelve, which covers related forming processes.

FIVE
Cooling of Thermoformed Parts

COOLING THE PART after forming is as vital as heating but can be even more time-consuming than heating. Therefore, it becomes important to select the appropriate method for it. Sometimes, when forming heavy-gauge parts, which tolerate the least internal stresses, it might be advisable to retard natural cooling by covering the formed part with a soft blanket. To prevent the buildup of high stresses in heavy-gauge parts, it might become also advisable to slightly loosen the clamps holding the sheet during cooling, even though additional shrinkage may take place.

Most of the heat absorbed during the heating cycle must be extracted from the plastic part before it can be removed from the mold. Otherwise, distortions or warpage in the part may occur. If the part is formed over a male mold, it must be extracted from the mold before it strangles the mold due to thermal shrinkage making removal too difficult.

Means of cooling the formed part

Conduction and convection are the only practical means of heat removal. The temperature of the formed sheet is too low for the dissipation of heat due to radiation. Because the heat conductivity of plastics is low, long cooling times must be accepted for parts having a wall thickness in excess of 1/16 in. If only a few parts are to be made, a draft of air from a fan will facilitate cooling, but care must be exercised to ensure uniform and even cooling. This method has the advantage that the sheet-clamping frame will be cooled at the same time. The disadvantages are that it cannot cool the mold sufficiently and that it must be timed so as not to interfere with the following heating cycle. Draft shields must be employed when cooling of parts and the heating of the next sheet take place together in the same vicinity.

There are significant differences between the cooling process for roll-fed automatic thermoformers and larger heavy sheet-forming equipment. Cooling the part in contact with the mold is very efficient if highly conductive metal molds containing water cooling channels are used or at least are mounted on cooled mold bases. By means of turbulent coolant flow it is possible to maintain a mold temperature within a $\Delta 5°F$ span. By keeping the cooling water temperature high, close to the plastic solidification temperature, the formation of chill marks—usually recognized by wave-shaped surfaces or abrupt steps in wall thickness—can be prevented. This can only be accomplished if thermostatically controlled chillers with heat exchangers are employed. The application of mold heat sink compounds will improve heat transfer considerably. The minuscule air space between mold and base becomes thus replaced by a compound consisting of zinc oxide in silicone oil.

Even if just a few prototype test pieces are being made on metal molds, which are not provided with temperature controls, the metal molds should be preheated either with the available sheet heaters or with a propane torch. Similarly, if the molds should get too hot, their temperature could easily be lowered below the boiling point of water by wiping them with a wet sponge.

When mineral- or metal powder–filled plastic molds or other water-cooled molds having low thermal conductivity are used on sheet-fed thermoformers, the temperature of the formed sheet drops at the mold side due to raising the mold's temperature and dissipating heat on the other side by air convection. Thus, the mold temperature becomes subject to wide temperature swings due to the lag in internal heat flow. By regulating the direct cold water flow or the rate of production, adequate cooling can be obtained. Wooden molds—due to their heat buildup—are not suitable for fast production runs.

Those parts that are produced according to the free forming or bubble process as described in Chapter Ten require the longest time for cooling because heat dissipation is limited to natural air convection.

To comprehend the importance of mold cooling, the example from page 17 should be used again; this example established that 55 kilowatts or 3100 Btu/min were required to heat the 10-mil-thick polyethylene sheet (42" wide, running at 100 ft/min) from 70 to 350°F. We assume that a mold temperature of 150°F has been chosen, but the part is removed from the mold at 160°F. Because the forming temperature was 350°F, the temperature difference is $\Delta 190°F$. Because polyethylene will crystallize during cooling, the heat of fusion, 33 Btu/lb must be added:

$$1200 \frac{\text{in.}}{\text{min}} \times 42 \text{ in.} \times 0.01 \text{ in.} \times 0.033 \frac{\text{lb}}{\text{cu in.}}$$

$$\times \left(0.55 \frac{\text{Btu}}{\text{lb °F}} \times \Delta 190°F + 33 \frac{\text{Btu}}{\text{lb}} \right) = 2290 \frac{\text{Btu}}{\text{min}}$$

This corresponds to approximately 10 standard tons of refrigeration. To obtain a uniform mold surface temperature, the water from the chiller is maintained at a constant 135°F temperature with a return temperature of 140°F. The amount of water to be circulated through the mold is:

$$2290 \frac{\text{Btu}}{\text{min}} \times \frac{1}{\Delta 5°\text{F}} \times \frac{1 \text{ lb}°\text{F}}{1 \text{ Btu}} \times \frac{1 \text{ gal}}{8.34 \text{ lb}} = 55 \frac{\text{gal}}{\text{min}}$$

If such a high flow rate cannot be obtained with the equipment at hand, one might have to subdivide the flow via a larger number of manifolds. Another possibility would be, if a wider temperature span must be chosen, to alternate cooling channels for the incoming and outgoing coolant. The water-cooling channels for the mold must be at least 1/2". The importance of the proper layout of these channels is understandable if one realizes that mold temperature fluctuations for high-density polyethylene should stay within ±2°F to obtain acceptable dimensional reproducibility over long production runs.

For engineering plastics, which have a much higher setting temperature (exceeding the boiling point of water), the coolant water must be kept at higher pressures or other heat transfer fluids must be selected. But none of them can attain the high heat removal rate of water.

The speed of heat withdrawal is dictated by thermal diffusivity, which is simply defined and can be determined empirically (see page 118). Because experience shows that approximately 0.005" of film thickness requires 1 second cooling time, the mold must be 40" long by 42" wide for the above example. This illustrates that cooling is often the factor governing production time. A smaller size mold would have required a reduction in running speed.

Observed cooling times can sometimes run counter to the expected values, derived from tests of similarly sized parts. Once the intimate contact between metal mold and the plastic is severed, heat transfer is markedly reduced. Limited duration of pressure, female cavities, and highly crystalline materials (such as polypropylene) represent circumstances that impede cooling rates.

In some instances it may pay to use a higher priced plastic that requires higher forming temperatures, such as polycarbonate, because its accelerated cooling time will allow an increased production rate.

Furthermore, on roll-fed thermoformers, when during the time span of the forming cycle only one cooling cycle takes place, several heating cycles are usually lined up with the help of multiple heater stations.

The sizing of the mold has been described in Chapter Three under part shrinkage. It is important to realize that the temperature of the mold and not the forming temperature will dictate mold shrinkage. The higher the mold temperature, the more shrinkage can be expected.

Non-conventional cooling methods

Faster cooling methods, such as a fine spray of water mist or of liquid carbon dioxide, can accomplish cooling rates that are tenfold the rate of forced air convection or hundredfold the rate of just ambient air convection. These processes are untidy or costly and are applied infrequently. Either might be justified, especially if applied locally to prevent hot tearing of highly drawn parts. It remains true that rapid and irregular cooling of formed parts does induce stresses that could impair their durability over time or can cause warping.

In a few instances the cooling cycle must be delayed or extended when materials are used, which require some dwell time at elevated temperatures to either complete a cross-linking reaction or accomplish crystallization to obtain improved mechanical properties (see crystallizable polyethylene terephthalate).

SIX
Trimming of Thermoformed Parts

ALTHOUGH FLASH REMOVAL of compression-molded parts and degating of injection molded parts can sometimes pose problems, the selection of the proper trimming method for thermoformed parts is an even more formidable task.

Most thermoforming processes start with expensive material, an extruded or calendered sheet; therefore, trim losses should be kept to a minimum. Nonrectangular parts especially require a wide margin at the periphery, leading to increased trim losses. Square or rectangular shapes are more favorable in their usage of material than round ones. If round molds are staggered at 60° triangular pitch, the amount of trim can be reduced by 10% (see Figure 7.17 page 95). Only if an in-line extruder is available will continuous regrinding and recycling of excess material reduce losses significantly. Still the reworking of materials involves costs associated specifically with it.

There is no single, immutable time for trimming. Often, a smoother cut can be obtained when the material is still warm. Electrically heated dies can be used in special cases, but low-heat conductivity of the plastics requires some heat-up time. High-production parts are frequently trimmed in the forming mold. When stacking is not immediately performed at the machine, multicavity formed parts should be left connected to the trim skeleton with a few narrow links to prevent that one part accidentally remains stuck to the mold.

When parts are formed, which must match in size with metal parts, trimming and stacking are mostly performed in another piece of equipment. As a result parts can first be completely cooled to room temperature. The importance of monitoring the shrinkage of the formed part was discussed in Chapter Three.

Trimming can be accomplished in many different ways, originating on many disparate separation principles. Ductile separation can occur when compressive or shear forces are applied. The force to be exerted is roughly proportional to the area circumscribed in the stress-strain diagram. Brittle fracture, requiring generally much less force, takes place in most abrasion and saw cutting, including routering and nibbling processes. Thermal methods of separation are employed with hot wire, hot gas jet, and laser beam cutting and, maybe, water jet cutting can be included here. General rules are difficult to establish because each plastic material and each thickness range may favor a different method. In many cases the avoidance of disadvantages will have the greatest effect on making the best selection.

Even the smoothest mechanical separation will lead to the generation of trim dust, which can consist of minute brittle fractured particles, fiber-like hair fuzz, or small fused particles. Wide kerf cutting will generate abundant dust. All such small particles are, due to the static electricity of plastics, difficult to remove and cannot be tolerated in medical and food packaging applications. To minimize dust problems, grinders and cutters should be provided with efficient exhaust. The parts can be treated with antistats if they do not interfere with the following process steps. Ionized air or a fine water mist can also prevent the clinging of particles to the plastic. The last resort is the washing of the parts with a mild detergent solution.

The thermal separation processes entail odor problems. Some disadvantages can be mitigated by applying the knowledge that the properties of plastics are very much temperature and rate dependent. Therefore, trimming is done sometimes at elevated temperatures to prevent fracturing and sometimes at ambient temperatures to obtain a smooth cleavage. The speed of power tools and feed rates must be selected to avoid heat generation and gumming. The guidelines for machining, published by the manufacturer in most data sheets for thermoplastics, should be consulted first.

Tools for trimming

The simplest tools represent sometimes specially shaped knives and scissors. Steel rule dies are convenient to manufacture and work very well with thin parts if all cutting is performed in one plane. They consist of hardened, sharpened strips of steel 1/16 in. wide and 1 to 2 in. tall. An arbor press, pneumatic cylinder, or a dinker machine working against a piece of end-grained hardwood blocks, hard rubber, or plastic pads may be sufficient. Foam rubber pads on the sides of the steel rules will strip thin materials off the cutters. This becomes especially useful when small cutouts are made at the same time. A low-cost machine that does the cutting of blister packs by passing them through two rolls requiring much lower pressure is shown in Figure 6.1. For heavier parts or for longer production runs the sturdier steel rule dies or forged dies become applicable.

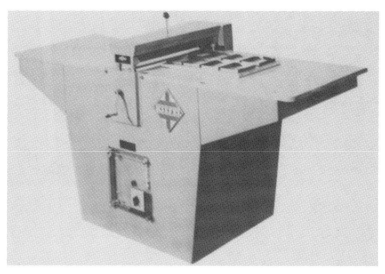

Figure 6.1. Cutting machine for blister packagings (courtesy of Paul Kiefel GmbH, D-83395 Freilassing, Germany.)

For high production, thin-gauge thermoforming punch-and-die trimming is employed. This can be done both as trim-in-place when the part is still on the mold as shown in Figure 6.2 or in-line or in-phase activated in a separate press (see Figure 7.10). They are expensive and need much care for correct adjustment. If pieces are to be trimmed horizontally, e.g., at the sides of the formed part, shear dies can move quietly in a planetary motion. In other cases the formed web is guided over a camelback or humpback and lowered vertically so that horizontally punched out parts can also be continuously stacked to assist in the counting and packaging of formed parts. Dies that work well for sheet metal parts may pose difficulties when used to shear plastics due to their excessive clearance. For plastics, practically no clearance should be provided. As can be seen in some of the illustrations in the next chapter, a heavier piece of equipment is required for trimming than for forming (Figure 7.10). One manufacturer's equipment uses a trimming die to cut the hot sheet partially before it is formed to a thickness down to 10 mils, regardless of the original thickness. This is still sufficient to keep the sheet in place during forming. When the material cools, final trimming is completed (Figure 6.2), requiring considerably less force.

To accomplish accurate trimming of thermoformed parts, it is necessary to provide in each part a certain area for accurate positioning and holding. Small thin-gauge parts are preferably held in position by surrounding interstitial locators. For parts having compound shapes or those

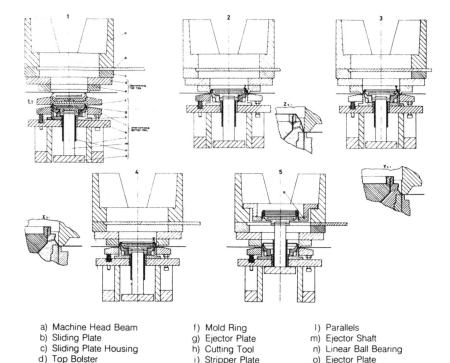

a) Machine Head Beam
b) Sliding Plate
c) Sliding Plate Housing
d) Top Bolster
e) Cutting Die
f) Mold Ring
g) Ejector Plate
h) Cutting Tool
i) Stripper Plate
k) Bottom Bolster
l) Parallels
m) Ejector Shaft
n) Linear Ball Bearing
o) Ejector Plate
p) Stacking Device

Figure 6.2. Forming and trimming tool. (1) Bottom tool being lifted. (2) Detail Z: sheet is held and partly sheared hot prior to forming. (3) Trays are pressure formed. (4) Detail X: sheet is completely sheared after forming. (5) Formed part is ejected and stacked (courtesy of Gabler Maschinenbau GmbH, D-23512 Luebeck, Germany).

that must be trimmed at different levels or angles trim fixtures should be made. They could consist of wood or glass fiber-reinforced plastics provided with metal edges. Generally, other part processing steps, which can be conducted quickly, are performed simultaneously with trimming, such as lip rolling, drilling, machining for cutouts or thickness tolerances, deburring, and up to snap-fitting parts together. The necessary templates and bushings can be included in the trim fixture.

The forces required for cutting thin plastic sheets up to 30 mil thickness with steel rule dies vary with the particular material. At a thickness of 10 mil, softer materials, such as the cellulosics and polyolefins, can be severed at cutting forces of 150 lb/in. of rule length. Tougher materials, e.g., the polyesters and oriented polystyrene, might require 500 lb/in. Most others, including polyvinyl chloride, fall in between at 250 lb/in.

Shearing and blanking of cellulosics and other sheets—even very thick ones—can be performed at room temperature, whereas polymethyl methacrylate sheet should generally be heated above 200°F to prevent the formation of star cracks. The published shear strength values for plastics do not indicate the required forces accurately, because too much depends on sheet thickness, accuracy and sharpness of the tooling, and the required quality of the cut.

An electrically heated wire can be used to advantage for separating thick plastic foam parts, resulting in a very smooth edge.

Large pieces, for which dies would be too expensive, must be manually trimmed or with programmed automatic tools. Nibblers represent an ideal tool because they do not generate many dust particles. However, they have not obtained the popularity that routers and saws have gained. Sometimes, it might become necessary to debur the cut edges. This can be accomplished either with a sharp edged scraping bar, a file, fine sanding paper up to buffing wheels.

With the introduction of robots in secondary processing, the three to six axis robots are enjoying great popularity for manipulating trimming tools, such as routers, laser beams, and water jet trimmers. In the latter case a fine stream of water impacting the sheet at a speed of 3000 feet/sec under 50,000 psi pressure severs the plastic much faster and more accurately (1 mil) than any other conventional method. Trimming can also entail center cutouts with no further finishing required (Figure 7.7). A catching device should be used at the opposite side of the sheet to muffle the high-pitched sound and collect the small amount of water and wet dust. For tougher and mineral or glass fiber-filled materials a fine abrasive is added to the water. But this also causes some wear in the equipment.

Automated equipment that cuts by means of a laser beam results in a still finer cut of superior precision. A 50-W CO_2 laser may cut through 1/2" of acrylic sheet at a speed of 1 ft/min. The edges will look as if they had been repolished. Because cut edges have been exposed to high temperatures, the part should be checked for heat damage and environmental stress-crack resistance. If necessary, parts should be annealed afterward.

SEVEN
Thermoforming Equipment

IN THE PRECEDING chapters the various components involved in the overall thermoforming process were described. Due to the great variety of processes, thermoforming equipment may assume many dissimilar shapes. At one extreme, hardly any equipment is required. Aside from an oven, makeshift molds alone may satisfy all requirements for a hands-on operation in which only a limited number of simple curvature parts have to be produced. At the other extreme, every 2 seconds a fully automated thermoformer, as shown later in Figures 7.23 and 7.24, can start with pellets and shape multiple parts occupying an area up to 21" by 13".

The material—if supplied in sheet form—may be fed manually or lifted by suction arms to the forming machine where it is clamped in a metal frame. Larger and heavier sheets may first be heated in any of the described heating ovens, either while suspended or lying flat. Medium-sized sheets are conveyed into the oven clamped in a frame. Ovens with infrared heaters can temporarily be slid over the frame. On the other hand, the frame with the sheet is usually pushed in to the oven when sandwich heaters are employed. If the material is taken from rolls, it is essential that a dancer unwinder is utilized to convert a gentle unwinding rotation of the heavy roll into the sudden movement required for the advancement of a new section of the sheet to be formed. The edges of the sheet are grabbed by pins or wedge-shaped teeth attached to chains, on both sides of the sheet. The chains, which glide in rails, transport the sheet through the heated tunnel ovens, the form press area, and sometimes also the trim section if it is part of the machine. Depending on material and machine layout temperature control (heating or cooling) must be provided to keep the plastic in grip. The endless chain returns for picking up new material. Electric servomotors, which provide a smoother and better controllable indexing motion, have replaced the older type mechanical rack-and-pinion arrangements.

To obtain high productivity, the heater tunnel is divided into two to four sections, each identical in length to the indexing stroke. Temperature zones are usually laid out in receding order. Heat profiling does not only occur in the indexing direction but also in the transverse direction to compensate for heat losses at the sides.

The actual forming process occurs in the main part of the machine. This area is equipped with one or two stages to carry the molds and the plugs on either side. These stages are movable in the vertical direction and must carry the vacuum or pressure lines as well as the coolant lines. At least one vertical motion between the mold and the sheet frame is necessary to create an airtight seal between them, so that the atmospheric air pressure (in the case in which a partial vacuum is applied) or the compressed air can shape the sheet. The second movable stage is frequently employed for more complicated parts, where a plug will first preform the sheet to obtain a better balanced material thickness during the final forming step.

Forming processes requiring only a one-sided mold are augmented by the matched-mold forming processes derived from metal stamping. Two mold halves with the right clearance to accommodate the thickness of the plastic material are mounted opposite each other and are closed over the heated sheet either by pneumatic or mechanical means. Although vacuum or air pressure is seldom applied in the mold, adequate venting must be provided. The forming process is, in all cases, quite rapid, but the material must remain in that section of the equipment until much of the heat from the plastic is removed by one of the cooling methods described.

Equipment for the mass production of smaller parts will invariably have trimming, stacking, and packaging capabilities in-line. Large pieces and low-production parts are usually trimmed, inspected, and finished away from the thermoforming equipment.

Single-station thermoformer

The least sophisticated equipment has only one work station. There, the sheet is clamped firmly and conveyed into the heater. The heater may also be positioned above and/or under the sheet. Usually, the mold is raised, and sometimes a plug is lowered into the sheet. After forming, the part is cooled while still on the mold. Figure 7.1 shows a small laboratory thermoformer suitable for making sample pieces, such as drape-formed parts or skin packagings. In Figure 7.2 a larger thermoformer is shown tooled up for a pressure drape-forming process. The mold is mounted to the top platen, which is lowered into the sagging heated sheet. In Figure 7.3 a modular design approach is illustrated. Flexibility with machines is greatest where they can be converted to the production of distinctly different parts.

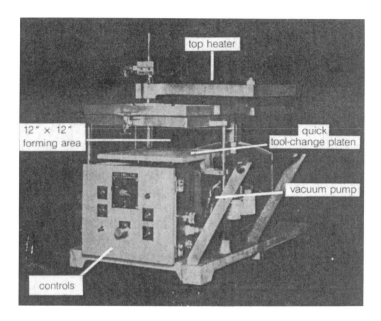

Figure 7.1. Laboratory sampler thermoformer (courtesy of Atlas Vac Machine, Cincinnati, OH 45242).

Figure 7.2. Single-station pressure former.

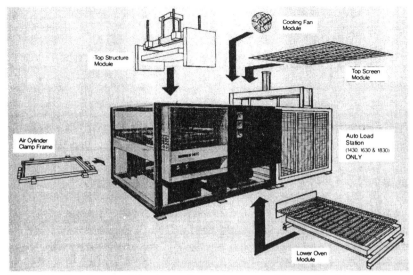

Figure 7.3. Cut sheet modular thermoformer (courtesy of Brown Machine, Beaverton, MD 48612).

Shuttle thermoformer

Shuttle-forming equipment virtually doubles output and conserves energy consumed by the heaters. This necessitates a second mold and forming station and also two clamping frames, which can be shuttled in succession from the oven to mold station A and from the oven to mold station B, respectively. A double-ended thermoformer as shown in Figure 7.4 can be operated with two crews producing two different parts. An infrared sandwich heater occupies the encased space between the two workstations.

Rotary thermoforming equipment

Greater productivity can be generated if three or four workstations are arranged around a central point. Only one mold and one forming station are needed. Three or four clamping frames mounted on a horizontal wheel and a means for rotating them from one station to the next are required, compared with the single-stage formers. An operator must be present only at the first station, where the frame is opened, the formed part removed, and a new sheet inserted. Station two provides heat to soften the plastic, whereas station three is the site of both thermoforming and cooling. The fourth station was originally added to accommodate a second heating oven for forming heavy-walled parts that require an extended heating cycle. Subsequently, the fourth station was utilized for automatic or robotic trimming. Again, the sheet heating cycle will dictate production speed. Because all other steps are performed during the same time as the heating cycle, good production efficiencies are achievable.

One such machine is shown in several of the following illustrations. Figure 7.5 is a drawing and a plan view of a four-station rotary thermoformer. Both top and bottom heaters can be subdivided, and the bottom heater accommodated for a downward motion to follow the sag of the

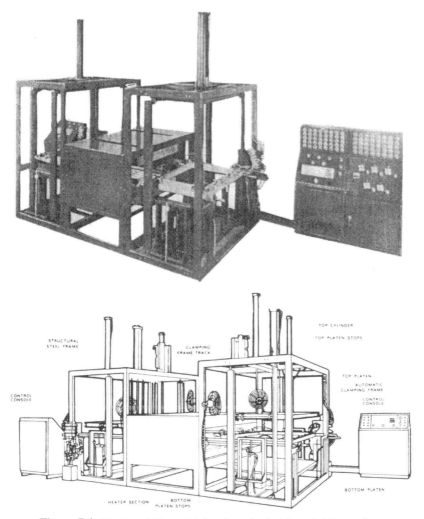

Figure 7.4. Photo and illustrated drawing of a double ended thermoformer.

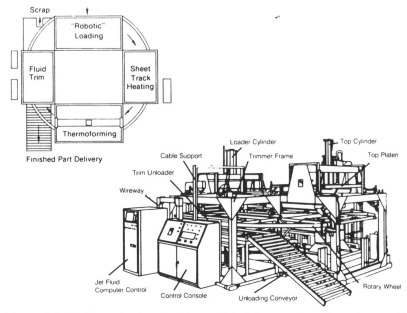

Figure 7.5. Drawings showing the "Thermoldtrim" System rotary thermoformer with fluid-jet cutter.

Figure 7.6. Photograph showing the rotary thermoformer sketched in Figure 7.5.

Figure 7.7. Close-up view of a fluid-jet cutter mounted on a rotary thermoformer.

heated material. Thermoforming could be performed both by vacuum and pressure forming. Figure 7.6 gives a side view with the fluid trim station in front with a close-up view in Figure 7.7. Details of this trimming process were described in Chapter 6, and the picture of the ring wheel assembly of a four-station, twin-sheet pressure former are shown in Figure 10.13. The largest rotary former by Brown Machine can use 10 × 20 ft sheets.

Large rotary thermoformers are sometimes too slow for the movement of the heated sheet from the oven station to the forming station. Thinner gauge or higher temperature plastics, such as polycarbonates, cannot be overheated. But they approach the low temperature point of their thermoformability very rapidly after removal from the oven. Such plastics are better formed on equipment with movable heaters or with heaters positioned directly over the forming table (canopy-type heaters).

Continuous in-line thermoformers

Fully automatic equipment utilizes a continuous roll of film. All machine functions are stationary and take place simultaneously. The film is advanced linearly from station to station at predetermined intervals. Because heating (and to some extent cooling) takes up appreciably more time than the other operations, the film is conveyed through a heating tunnel containing from 4 to 10 heating stations. This entails that the cooling cycle becomes the dominant factor in the overall process.

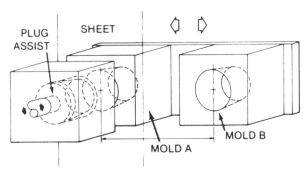

Figure 7.8. Horizontal shuttle mold machine. Sheet travels vertically downward. Molds shuttle left to right. Forming station remains stationary (courtesy of *Plastics Machinery & Equipment*, Dec., 1978).

For thin films, efficient cooling can take place while the plastic remains on the mold. In other cases, after adequate rigidity is attained to allow removal of the part from the mold, final cooling is carried out down line by forced air convection.

The output of formed parts can be nearly doubled when two cooling cycles in a row are utilized. This may be especially necessary when working with materials of low thermal diffusivity, such as polypropylene.

The previously mentioned shuttle-mold principle has, therefore, been applied also to continuous thermoformers by utilizing horizontally or vertically reciprocating mold pairs. As shown in Figures 7.8 and 7.9, the heated sheet travels vertically downward (Z direction), whereas the two

Figure 7.9. High-speed thermoformer using horizontal shuttle molds as sketched on Figure 7.8 (courtesy of *Plastics Machinery & Equipment,* 1978).

molds shuttle horizontally in the X direction. Usually, a plug assist will push the sheet horizontally (Y direction) into the mold. Compressed air completes the forming, and a trim die separates the parts before the molds can be shuttled again. While forming takes place in mold A, the previously formed part cools in mold B.

Most automatic thermoformers have an intermittent film-feed mechanism. During all working cycles the film remains motionless. The web is quickly indexed to the next position after completion of the forming step. A multitude of drive systems are employed to obtain smooth acceleration and deceleration, which cut down noise and wear and speed up production. Adjustments can usually be made while the machine is working, and optional equipment may photo scan preprinted films to obtain perfect register.

Several types of continuous thermoformers are depicted in the following illustrations. Figure 7.10 shows a 25-ton forming press with a counting stacker. The corresponding drawing in Figure 7.11 outlines the various machine components and their functions. A photo of a less sophisticated machine with similar capabilities is shown in Figure 7.12. The size of the directly acting air cylinders clearly shows that the power requirement for the trimming station is higher than for the forming station.

The photo in Figure 7.13 depicts the Model 44 Thermoformer, provided with a double-length heat tunnel. Figure 7.14 is a close-up of the mechanical drive of the former with the safety guards opened for better visibility. This machine could be considered the smoothest running thermoformer. Its MicroPhaser II control regulates the entire manufac-

Figure 7.10. Electropneumatic 25-ton thermoformer.

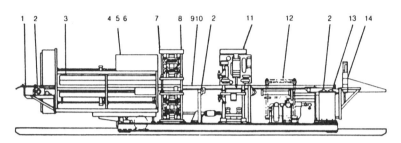

Figure 7.11. Same thermoformer as in Figure 7.10. Captions illustrating functions of machine components.

Figure 7.12. High-speed preheat pressure former with two stations, safety guards removed for illustrative purposes (courtesy of Atlas Vac Machine, Cincinnati, OH 45242).

Figure 7.13. Model 44 Thermoformer (courtesy of Irwin Research & Development, Inc., Yakima, WA 98902).

Figure 7.14. Mechanical drive linkage of thermoformer shown in Figure 7.13 (courtesy of Irwin Research & Development, Inc., Yakima, WA 98902).

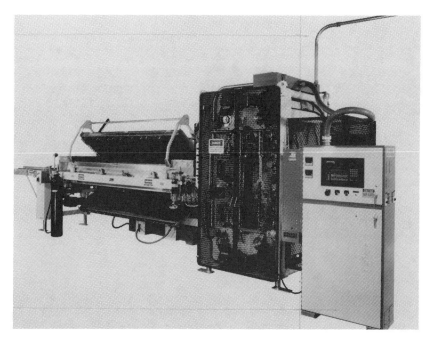

Figure 7.15. Opened 16 zone heat tunnel of thermoformer shown in Figure 7.13 (courtesy of Irwin Research & Development, Inc, Yakima, WA 98902).

turing process. Both the thermoformer and the trim press are automatically synchronized at operating speeds between 1 and 145 cpm, with the capability of making adjustments on-the-fly. Functions can be varied, in any sequence, with the automatic correction of other interrelated actions assured. All motions are executed by brushless servomotors with rotors containing lowest inertia rare-earth permanent magnets. The motors are controlled by pulse width–modulated (PWM) transistor technology in place of the older, inferior SCR controls. A close-up of the infrared heaters in an opened position is found in Figure 7.15 showing a smaller, 28″ mold width machine. The specifications for Model 44 Thermoformer are listed here as representative of machinery data. This type of information is usually provided in most manufacturers' equipment brochures but cannot be cited at all points in this book.

Material Index
Sheet width
    Maximum 46″
    Minimum 30″
Maximum chain speed
    141″/sec

Automatic chain rail adjust (optional)
MP II precisely controls the chain rail positioning facilitating immediate changes in sheet width. A gear box and motor operate each of the adjust points. The rails can also be automatically moved into a V shape, either narrower or wider at the former, to compensate for severe sag or orientation in the sheet.

Heat Tunnel
Heat tunnel length
80"–200"
Heating elements
Calrod Standard. Ceramic and quartz optional.
PID Heat control (optional)
PID Heat individually controls up to 48 thermocouples and 84 heat zones. Heaters can be assigned to different thermocouples to match the heat zone layout to the product. Tunnel length and width can also be adjusted by turning off heaters as part of the product recipe. All parameters are part of the product recipe, stored on hard disk for fast access. MP II's optional video control shows configuration of the tunnel, all temperature setpoints, and all actual temperatures.
Temperature variance
With PID control, typically no more than ±2 degrees at 600 degrees Farenheit (.66% within any given zone).

Former
Maximum mold size
44" Wide
40" Long
Depth of draw
Model 44: Maximum 2.50" above or below sheet line
Model 44 Mini-Mag: Maximum 5.50" above or below sheet line
Former press rating
80,000 pounds
Press opening in closed position
Model 44 standard shut height is 6.000" (3.500" from sheet line to top platen, 2.500" sheet line to bottom platen).Optional shut heights up to 10.00" (5.000" sheet to top platen, 5.000" sheet to bottom platen) available upon request.
Model 44 Mini-Mag maximum shut height is 15" sheet line to top platen, 10" sheet line to bottom platen.

Platen travel
  Model 44: 3.78" per platen (at 6" closed height)
  Model 44 Mini-Mag: 8.50" per platen (at 20" closed height)

Other Specifications
Machine control logic
  Machine control functions are accomplished via the Micro-Phaser II system
Machine layout
  Right or left hand model
Machine color
  IRAD blue standard. Other colors availble upon request.
Quick change tooling system (optional)
  The mold is held in place by clamps which hold the tools with spring pressure and are released by hydraulic pressure. The tool is held by four clamps on top and four clamps on bottom. A common hydraulic pump operates all clamps. Tools can be simply modified to accept the clamp by adding threaded inserts to each tool.

Modern Plastics Encyclopedia contains similar data for many thermoformers sold in the United States.

Finally, Figure 7.16 shows the camelback arrangement of the Model 50 Trim Press, which would be positioned to the right of the thermoformer shown in Figure 7.13. For all control functions it utilizes the MicroPhaser II control of that equipment.

In mass producing parts, speed of operation and optimal utilization of material are of utmost importance. The photo in Figure 7.17 shows an automatic thermoformer that is also sketched in Figure 7.18. After forming and trimming, the female mold is tilted 80° for the combined part ejection and stacking stroke. The 30° arrangement of the round molds results in the best material utilization and easiest trimming.

A 28-cup tool for a similar automatic thermoformer is shown in Figure 7.19. The tool is moved upward in a horizontal position for forming and trimming and then lowered and tilted 80° for the ejection and stacking of molded cups.

Another possibility for improving production speed is sketched in Figure 7.20, where four sets of molds are arranged crosswise, multiplying the time available for cooling by 3. On the payoff side, heated rolls dry the film material. The grinder on the right hand side converts the trim web into recyclable granulate. The magnified cutout in Figure 7.21 shows that plug assist forming occurs in position 1. In position 2 the lips can be rounded off. Position 3 serves only for additional cooling, and in position 4 the cups are ejected and stacked in a single stroke.

Figure 7.16. Model 50 trim press (courtesy of Irwin Research & Development, Inc., Yakima, WA 98902).

Figure 7.17. High speed automatic thermoformer and stacker (courtesy of Adolf Illig Maschinenbau GmbH & Co., D-74080 Heilbronn, Germany).

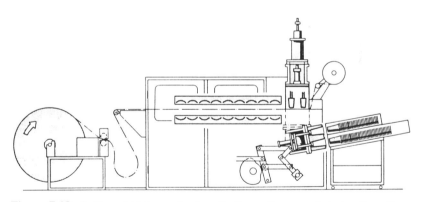

Figure 7.18. Outline of high speed automatic thermoformer and stacker from Figure 7.17 (courtesy of Adolf Illig Maschinenbau GmbH & Co., D-74080 Heilbronn, Germany).

Figure 7.19. The cup tool for the thermoformer M91 is shown with the dies tilted for servicing with the ejectors activated (courtesy of Gabler Maschinenbau GmbH, D-23568 Luebeck, Germany).

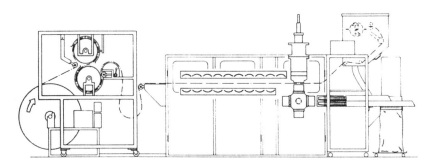

Figure 7.20. High speed thermoformer and stacker utilizing a quadruple mold (courtesy of Adolf Illig Maschinenbau GmbH & Co., D-74080 Heilbronn, Germany).

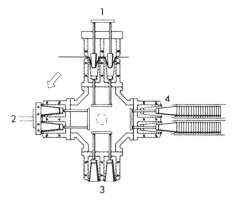

Figure 7.21. Enlarged detail of thermoformer from Figure 7.20 showing working steps of rotating quadruple mold (courtesy of Adolf Illig Maschinenbau GmbH & Co., D-74080 Heilbronn, Germany).

In-line thermoformer

If the forming process can be accomplished during the time it takes to extrude the sheet and if long production runs are involved, an in-line extruder thermoformer could be considered. The advantages are that the material is already available at a very uniform temperature and it might not have to be reheated. This is especially important when forming parts out of foamed sheet materials. Furthermore, any cutouts, margins, and trim materials can continually be reground and reextruded, thus solving the problem of eliminating generation of scrap material and reducing the likelihood of contamination of regrind. Because the regrind virgin resin ratio remains constant for each product, the disturbances found when external regrinds are reprocessed are avoided.

The problem of making the steady motion of the sheet extrusion compatible with the interrupted motion of the sheet in the thermoformer has not been fully overcome. If a traveling mold, which retracts after each forming, is chosen, vibrational disturbances may affect the quality of the sheet at the die lip. Therefore, in most cases, the sheet is still first partially cooled when passing through a roll stack and then reheated in the thermoformer. A loop in the sheet or a pair of dancer rolls between both machines render the two different transportation modes compatible.

The changeover to in-line processing has been accelerated by the recent availability of well-tuned equipment components, all supplied by the same manufacturer. Improvements in extruder technology and the increased use of gear pumps, which eliminate extrusion surging, have helped too. The remaining disadvantages are that problems arising at any point may require the shutdown of the whole production line and that no preprinting of the sheet is possible.

Figure 7.22. In-line thermoforming system forming polystyrene cups.

Figure 7.22 shows such an in-line thermoforming system producing cups out of general purpose crystal polystyrene. The schematic of this machine combination is seen in Figure 7.23. and again for illustrative purposes the technical data taken from the manufacturer's brochure are reproduced here:

Extruder
- DC main drive motor with tachometer feedback
- 2 speed gearbox (single speed also available)
- Barrel is manufactured from heat-treated special steel and has a conical shaped vent hole (a plug is supplied for closing the vent).
- Liquid barrel cooling: The cooling system is fitted with pressure gauges and thermometers for precise temperature control.
- Venting unit is equipped with liquid cooled vacuum pump, with flowmeter to protect pump motor, condensate collection, and condensate gas gathering.
- automatic screen changer
- flexible lip die with restricter bar and adjustable deckles; sheet die cart

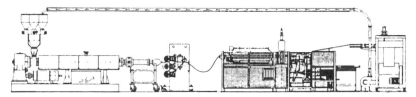

Figure 7.23. Schematic of in-line thermoforming system shown in Figure 7.22.

Roll Stack
- three-roll stack calender: upper and lower rolls adjustable
- DC drive with reduction gears
- rolls ground, hardchrome plated and super finished
- pull rolls with independent drive
- roll stack temperature control, each roll independently controlled

Fully Automatic Thermoforming Machine
- DC motor drive
- mechanically operated
- pointed chain sheet transport; infinitely adjustable index length
- sheet edge preheating
- ceramic oven heaters; SCR controlled with TC feedback
- independent timing of trim operation
- forming without trim during start-up mode
- programmable sequence control
- patented vacuum plate takeoff system
- mechanically controlled automatic stacking and counting

Scrap Recycling and Feeding
- grinder for skeleton webb
- pneumatic feeder for virgin material
- spiral feeder for recycled material
- hopper drawer magnet
- variable ratio, three-component hopper loader: virgin, recycle, color concentrate (on request a unit for 4 raw material components)

Technical Specifications

Material	HIPS, ABS (optional screw for PP)
Throughput	HIPS 880–990 lb, ABS 770–880 lb, PP 550–650 lb
Die opening Min-Max	HIPS 0.2–2.2 mm, ABS 0.2–2.2 mm, PP 0.4–2.0
Max. mold size	540 mm (21 1/4″) × 330 mm (13″)
Max. depth of draw	130 mm (5.118″)
Trim force	20 metric ton
Max. production speed	35 fpm
Connecting load (without motors)	231 KW total line
Connecting load motors	184 KW total line
Max. air consumption	4000 liter/min. at 6 kg/cm^2 (1056 gal/min at 80 psi)

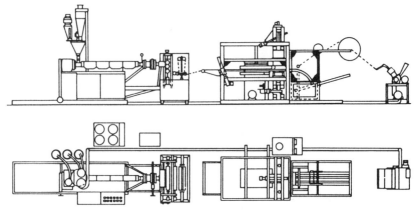

Figure 7.24. In-line thermoforming system for solid and foamed sheet parts (Kiefel Extrumat (courtesy of *Plastics Machinery & Equipment,* July 1985).

A comparable setup is shown in Figure 7.24. This Kiefel Extrumat in-line system is designed for converting both solid and foamed sheet to formed articles.

Linear thermoformers

There are, however, other types of automatic feed thermoforming machines in which the sheet keeps moving at a steady speed instead of the usual start-stop mode. They can be fed either from rolls with the film first traveling through a heater tunnel or from a film emerging directly from an extruder. In one case the set of molds, the clamps, the plugs, and all vacuum connections travel in unison with the sheet on a moving conveyor and return in a loop when the forming is completed. The set of molds could also be arranged firmly at the circumference of a drum. In both cases the need of multiple molds restricts such formers to large volume productions.

Figure 7.25 shows the schematic of a simplified continuous linear thermoformer by the Linear Form Pty. Ltd. A simple conveyor with 18 molds proceeds at the linear speed of 69 ft/min, dictated by a film or sheet extruder or a film-heating tunnel. A second conveyor, driven by the same motor, transports the heated film and preforms the part by means of a cam-actuated protruding plug. Final forming takes place by means of a blast of air. The need for leak-prone rotating manifolds for mold-cooling water is circumvented by the use of chilled air for part and mold cooling.

A new rotary thermoforming machine from Irwin International (Yakima, WA) is shown in Figure 7.26. In this case 10 sets of molds are

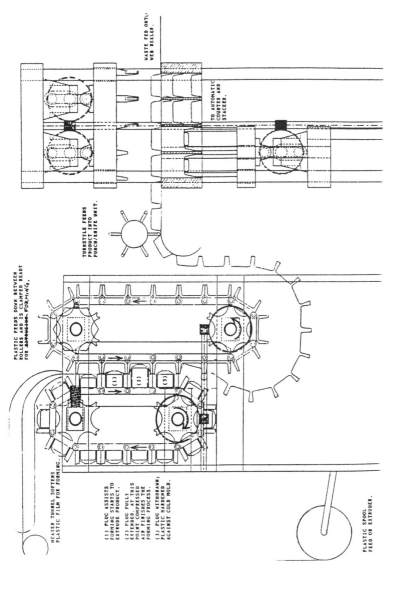

Figure 7.25. Drawing of the linear thermoformer shows the two endless chains to which the clampdown and plug mechanism on the left and the molds to the right are attached (courtesy of Linear Form Pty. Ltd., Brown Plains, Qld. 4118, Australia).

Figure 7.26. An in-line continuous rotary thermoforming drum is directly fed by a sheet extruder to provide 71,000 container lids per hour (courtesy of Irwin International, Yakima, WA 98902).

mounted on a continuously revolving drum rotating at approximately 10 rpm. The film or sheet material is supplied at the forming temperature directly above the rotor from the extruder. This arrangement obviates the necessity for providing heating tunnels and saves energy; however, extruder output must be tightly controlled to the roll former. Because molds for deep drawn parts would squander much material at the edges, the depth of formed parts is limited to 1 in.

Pneumatic thermoformers

A number of different motions must be performed by thermoforming equipment. Besides the transportation of the sheet web, the movements of the molds, the plug assists, the in-mold trim die, and the stripper plate must all be well coordinated. These motions can be accomplished by pneumatic, hydraulic, or mechanical means and often by a combination of them.

Pneumatic thermoformers were popular because they are least complicated and very versatile. They are fast and safe to operate because the moving parts are lightweight and the forces limited. Simple mechanical stops can quickly be adjusted for the right length of travel. Another advantage is that for food packaging, the possibility of contamination by oil is minimal. The limitations include the facts that most types are not built sturdily enough to process parts by the million and that they require more energy to accomplish all motions. A large, power-consuming air compressor must always run at peak pressure, even though a high percentage of this energy is wasted in performing low-pressure motions that, nevertheless, consume a high volume of air.

Hydraulically operated thermoformers

Hydraulic thermoformers are as versatile as, but more complicated and more expensive than, pneumatic thermoformers. One reason is that hydraulic actuators always need a return line. Hydraulic movements are smoother and more easily controllable, resulting in better accuracy. The power requirements are also lower. Hydraulic machinery used to be powerful but slow, noisy, and dirty. This is no longer the case. Better pumps and circuitry have overcome slow speed and noise, and improved valves, seals, and fittings have nearly eliminated oil weeping and leakage. Yet, more maintenance hours are required for hydraulic equipment, e.g., in spotting and repairing leaks, especially where stringent demands are set for food-packaging items. More care must be exercised to set up the various work functions in proper sequence. However, because these thermoformers are built more sturdily, they are better suited for long production runs. The power reserve of hydraulic thermoformers can be put to good use for postforming operations, such as trimming.

Mechanically operated thermoformers

Mechanical thermoformers were developed to exceed the productivity of any other thermoforming equipment. They represent the choice for "dedicated" thermoforming lines. The platen stroke depth must be fixed when a mechanical toggle or a crank action design is used. The shortest cycle times can be obtained in connection with independently powered DC servomotors. Web advancement with precise indexing—necessary when using preprinted stock—can also be accomplished with DC servomotor-powered pin chains. The most sophisticated controls are used with this equipment to utilize its full capabilities. Although this machinery may originally have been designed for a specific job and for a certain material, it is now available for many large-volume lines. The high capital investment is paralleled by the high-level technical capabilities of the personnel needed to set up and adjust the equipment. Mechanical thermoformers have the lowest frequency of maintenance requirements. Power consumption for the mechanical part of this equipment is negligible, whereas energy requirements for heating and cooling the plastic are dependent on bulk of output.

Skin packaging equipment

Thermoforming processes are not only employed to manufacture structural parts or packaging containers but have been specifically modified to incorporate packaging and sealing. For instance, with skin packaging, the items to be enclosed are placed on a piece of printed cardboard, which is air permeable and rests on the plenum chamber. After the plastic film is sufficiently heated, the mounting frame is lowered to obtain an air seal with the cardboard, and the vacuum is quickly applied. The softened sheet will stick to the specially coated cardboard and tightly adhere to the packaged goods. No molds are required for this process.

Blister packaging equipment

The blister packaging process requires blisters to be formed in female molds by a conventional thermoforming method. The pieces to be packaged are then placed into these cavities and are covered by a piece of cardboard that contains all information necessary for display. A heated pad must only momentarily be applied to the periphery of the plastic blister to heat seal it to the cardboard and to enclose the items. For smaller parts, multiple packages are usually sealed simultaneously, permitting the utilization of the machine's entire work area.

Snap packaging

In this process, usually two parts, namely, a container and a lid, are first thermoformed by any conventional method. After the container is filled with merchandise, the lid is snapped on, and a hermetic seal is obtained by the simple heat-sealing process described. For enhancing the freshness of perishable goods the remaining air in the container may be replaced by an inert gas just prior to sealing.

Vacuum packaging

Vacuum packaging can be combined with any of the three methods described. Plastic materials for both the base or container and for the closure must be airtight. For food packaging such plastics frequently consist of barrier polymer containing coextrusions or laminations. The material for at least the closure part must be semiflexible. After filling, the closure is placed over the container. The heat sealing is either performed in a vacuum chamber accommodating a number of packages, or the air is evacuated individually from each package prior to heat sealing. The trimming operation completes the process as in all other cases.

Packaging machinery

The machinery available for these various packaging processes is diverse, ranging from the simplest manual or foot-actuated to fully automatic machines. They all have in common that they are designed for accomplishing a wide variety of jobs. Figure 7.27 represents a small heat-sealing press for blister packages, and Figure 7.28 shows a small thermoformer for various packaging applications.

Two large production machines are shown in Figures 7.29 and 7.30. The latter is capable of vacuum and inert atmosphere packaging. The schematic of this process is sketched in Figure 7.31, and the setup for the

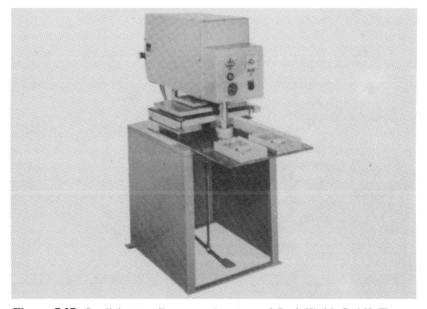

Figure 7.27. Small heat sealing press (courtesy of Paul Kiefel GmbH Thermoformmaschinen, D-83395 Freilassing, Germany).

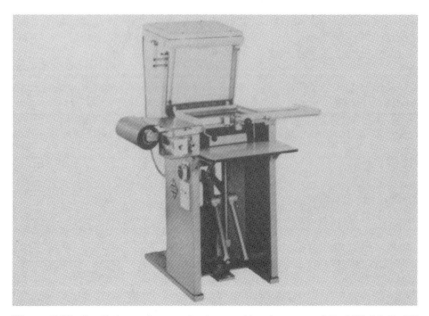

Figure 7.28. Small thermoform packaging machine (courtesy of Paul Kiefel GmbH Thermoformmaschinen, D-83395 Freilassing, Germany).

introduction of the inert atmosphere and the motions performed by the package container, the vacuum box, and the sealing film are pointed out in Figure 7.32. Figure 7.33 presents a close-up view of the equipment shown in Figure 7.30. The sealing chamber in which evacuation and flushing also take place is seen in the center. The photo in Figure 7.34 demonstrates the variety of jobs performed by this machine and the environment in which it is used.

Figure 7.29. Sureflow thermoform, fill and seal packaging machine system (courtesy of Mahaffy & Harder Engineering Company, Fairfield, NJ 07004-2914).

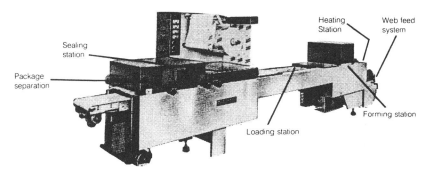

Figure 7.30. Tiromat VA automatic form, fill and seal machine (courtesy of Kraemer & Grebe GmbH & Co. KG, Maschinenfabrik, Biedenkopf-Wallau, Germany).

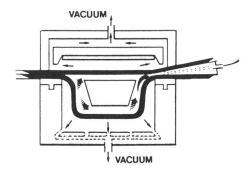

Figure 7.31. Schematic of vacuum packaging process.

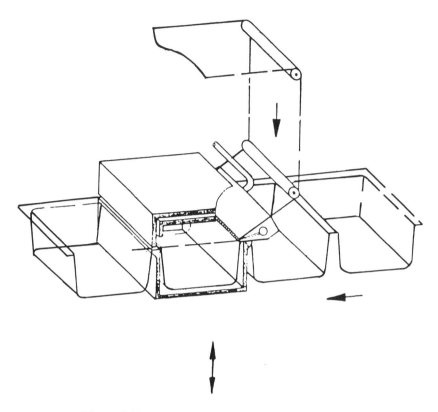

Figure 7.32. Layout for vacuum and gas flushing process.

Figure 7.33. Close-up view of Tiromat packaging machine with sealing chamber in center (courtesy of Kraemer & Grebe GmbH & Co. KG, Maschinenfabrik, Biedenkopf-Wallau, Germany).

Figure 7.34. Tiromat packaging machine in processed meat plant (courtesy of Kraemer & Grebe GmbH & Co. KG, Maschinenfabrik, Biedenkopf-Wallau, Germany).

Control mechanisms

One very important aspect of any thermoforming machine is its control mechanism. Any kind of equipment, from mechanical time-lapse relays to sophisticated microprocessors and computers, may be encountered on presently used equipment. Only the required control functions are briefly listed below.

Temperature controls are required for both the heaters and for mold cooling. Because the temperature sensors can never be located exactly at that point on the material where the temperature is to be controlled, and because time is required to transfer heat, there will always be a time lag between the sensor signal and the corresponding temperature adjustment. Simple on-off or timer controls have been replaced by proportional controllers, which resulted in a significant reduction in temperature fluctuations. Still, temperature changes persist due to variations in heat demand. Automatic reset controllers and newer automatic trimming circuits, aided by microprocessors, can keep the material temperature constant, even when the speed of automatic thermoformers is changed. Although all controllers will prevent the buildup of excessive temperatures, an independent temperature overshoot device is a good investment to avoid equipment damage in case of controller malfunctions.

Vacuum controls are used to maintain the accumulator tank vacuum. A separate vacuum gauge close to the mold table is always helpful for adjusting the forming process or when searching for causes of malfunctions caused by leakage.

Timing controls are crucial for continually controlling the repetitions of the heater, vacuum, and cooling cycle durations to establish the shortest sequencing of the various motions involved in the forming process.

Microprocessors and the—now at low cost available—personal computers have the advantage that they can be programmed to automatically and continually readjust certain processing conditions to correct for inevitable aberrations, such as gauge thickness, heater output, and other variations, detectable by sensors attached to the machine.

EIGHT
Thermoforming-Related Material Properties

BASICALLY, ALL THERMOPLASTIC materials should be suitable for thermoforming processes. Such materials, when heated, will exhibit a reduction in their modulus of elasticity, their stiffness, and their load-bearing capacity. To understand these relationships it becomes necessary to know how temperature changes affect the physical properties of plastics. We are too much accustomed to assume that our everyday materials, such as wood, concrete, glass, metals, and textiles, remain unchanged between 0 and 200°F.

Glass transition temperature

Low molecular weight or atomic crystalline solids will, upon raising the temperature, melt at a certain point by forming a low-viscosity liquid (water, salts, metals, and the like). The common metals become as fluid beyond their melting points as mercury is at room temperature. High molecular weight amorphous materials (most thermoplastics) or materials that exhibit strong ionic bonds, such as glass, can appear to be rigid solids as long as their chain links remain immobilized. Depending on their chemistry, these links may, upon heating, become at one point able to rotate and translate. This characteristic point has been named *glass transition temperature*. Above this temperature the material remains a coherent solid but exhibits some flexibility resembling a "leathery" or "rubbery" texture. Starting with an extremely high viscosity (in the range of 10^{12} poise[1]), the polymeric material will change with increasing temperature to a viscous substance but will never become a fluid liquid because the weak cohesive forces acting between each link of the long polymer chain will prevent the molecules from easily sliding past each other. At high temperatures exceeding 300 to 500°F thermal decomposition of organic materials will occur.

Semicrystalline polymers, depending on their degree of crystallinity,

---
[1] Water has the viscosity of 0.01 poise.

will exhibit still another transition at their crystalline melting point, which is usually much higher than the glass transition point.

Heat deflection temperature

The heat deflection temperature, a change in mechanical properties, represents a more practical temperature limit for the materials used in thermoforming processes. The literature usually lists two values. The first, determined under 264 psi loading, is the value for determining the temperature up to which rigidity for light mechanical load applications relevant to large parts is retained. The second value, determined at a quarter of that load, 66 psi, represents the upper temperature limit for applications to small, stubby parts.

This temperature, or in some cases a still higher temperature limit—sometimes called the no-load deflection temperature—is extremely significant for thermoforming, since the material temperature of the formed part must be below this temperature to be safely taken from the mold. Otherwise, gravitational force would collapse or distort the formed part. In view of this fact, the mold temperature setting should be at least 20°F below this temperature.

For many plastics the differences in loading have a minimal effect on the deflection temperature; however, for others they are significant. The lower loading became introduced when tests on nylon resulted in an unacceptably low value of 160°F for the heat deflection temperature versus 450°F, which was obtained when loading became reduced to just one-quarter of it (66 psi).

Softening range and hot strength

At still higher temperatures the material's behavior will again stabilize when exhibiting a rubbery state. In this temperature range a film or a sheet becomes easily stretchable but may also easily retract when the force is released. At increasingly higher temperatures the melting range of the thermoplastic material is reached, at which point it will gradually turn into a viscous state. This is observed in the rapid sagging of the heated sheet when the relatively low force of gravity becomes sufficient to cause deformation. In general, the usually broad, upper, semielastomeric range is most conducive for thermoforming.

Table 2.2 (page 28), which contains the properties of thermoplastic materials applicable to thermoforming, lists all relevant temperatures and the melting temperatures for crystalline polymers.

Softening range and hot strength are properties that will affect the suitability of plastics for thermoforming. A wide softening range, i.e., a broad temperature span in which the plastic is soft, pliable, and elastic, is desirable because during the forming process the temperature of the material tends to drop. A wide softening range provides the time span for intricate parts and for internal edges and corners that touch the mold last to be formed to the desired shape. If the material has insufficient hot strength, the heated sheet may rupture or thin out rapidly in certain spots due to the decreasing wall thickness in the stretched areas. This will lead to excessively thin areas or, in more severe cases, to punctures in the sheet.

In Figure 8.1 the changes in the *modulus of elasticity*, a numerical measurement to express the stiffness of a material, with increasing temperature are sketched for four materials, which exhibit about the same stiffness at room temperature. At a certain temperature each of them exhibit a brief range where the modulus of elasticity drops precipitously. For the amorphous resins the glass transition temperature can be found there. But for regular thermoforming processes only the succeeding range where the modulus of elasticity forms a viscoelastic plateau becomes important. There, small unavoidable changes or variations in temperature pose only a minor effect on the sheet's flexibility. Under those conditions these materials possess a viscous component, to permit the sheet to stretch under moderate forces, but also a sufficiently high elastic component, to resist flow just due to the much lower gravitational force. The much wider temperature range for the amorphous resins,

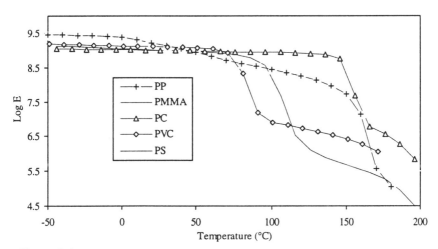

Figure 8.1. Temperature-dependent viscoelastic behavior of plastics obtained by dynamic mechanical thermal analysis (reprinted from N. Macauley, E. Harkin-Jones, and W. R. Murphy, The Queen's University of Belfast, SPE ANTEC Conference Proceedings 1996, p. 861).

polyvinyl chloride (PVC), polystyrene (PS), polymethyl methacrylate (PMMA), and polycarbonate (PC), is indicative for the much wider processing window for these materials. In comparison, the semicrystalline polypropylene (PP) has only a very narrow range. The height of the plateau becomes indicative for the force that has to be exerted during thermoforming; in this case PVC and PC are higher than the other plastics and polypropylene is the softest at forming temperatures.

To better understand the structural differences in these materials, one should visualize the two extremes: (1) a low molecular weight thermoplastic, having low intermolecular forces on one side and (2) a cured, highly cross-linked thermosetting resin, such as a cured phenol formaldehyde plastic, on the other. The first will rapidly turn into a decreasingly viscous liquid, having passed quickly through the elastomeric range. Thus, it exhibits a very narrow softening range and a low hot strength. The latter will remain quite rigid up to a temperature where thermal decomposition takes place. It has hardly any softening range and such a high hot strength, that thermal forming becomes highly restricted. Both kinds of plastics are unsuitable for thermoforming.

Therefore, plastics suitable for thermoforming have been specifically selected and developed over the years. Generally, much higher molecular weight thermoplastics are used for extrusion and calendering purposes than for other processes. Thus, the film and sheet materials normally fabricated by these processes contain a higher proportion of the higher molecular weight polymer. The opposite is true for injection molding and rotomolding processes. In these, lower molecular weight thermoplastics are used because the heated resin must have a sufficiently low viscosity to flow through the gates of the injection mold, or the granules must be able to coalesce (melt together) easily when rotomolded. In both cases the plastic will cool and solidify while its shape remains constrained in the mold.

In the extrusion process, the material will leave the die at its peak temperature, and the sheet will cool and solidify in the open, without the need for constraint. However, in most cases the sheet passes immediately thereafter through roll stands, not only to cool the sheet but also to improve gauge uniformity and surface evenness (gloss) or to impose surface texture. These are some of the reasons why thermoformed parts can be expected to produce parts with better mechanical properties than parts made of the same generic material by any other process.

Aside from high molecular weight, some tendencies at the molecular level will favorably affect the softening range and the hot strength. The arrangement of the elements in a ring structure will make the chain links more voluminous and less likely to slip past each other. The cellulosics, polycarbonates, and the aromatic (benzene ring containing) polyesters, among others, fall into this group. These ring structures will raise the stiffness of the polymer chain also when they are just present in the side chains, as can be seen when comparing the high modulus of elasticity of polystyrene (450,000 psi) with that of polyethylene (40,000 to 200,000 psi).

The high polarity of the chlorine atom in polyvinyl chloride is also very beneficial. One should remember that polyvinyl chloride is the only thermoplastic resin that can tolerate compounding with large amounts of plasticizer and still retain an extremely high creep resistance. Biaxially oriented film materials, such as biaxially oriented polystyrene, are also well suited, because a certain amount of slack within the chains has already been removed during orientation, comparable with a taut rope. Still another possibility for obtaining higher rigidity is to select crystalline materials (e.g., the polyolefins, the nylons, and especially the high-temperature polyphenylene sulfide and polyetheretherketones), whose chain mobility is restricted due to their partial fixation in a crystal structure. However, one should bear in mind that these forces disappear once the crystal melt temperature has been exceeded, thus often making crystallinity of questionable value.

The incorporation of minute numbers of cross-links between the chains of thermoplastic resins strengthens the heated plastic. Cross-linking may take place during polymerization due to the presence of trivalent monomers or to chemical reactions with cross-linking agents. Both the cell cast as well as the continuously cast acrylate sheets should be mentioned here. Because they contain some cross-links, a marked difference in formability between these sheets and extruded acrylic sheets is noted. Similar observations are made when large amounts of regrind are reprocessed for the extrusion of sheets. In this case cross-linking—mostly due to oxidation during thermal processing—increases the molecular weight. On the other hand, if a breakdown in molecular weight occurs during reprocessing because of hydrolysis, the material may become unsuitable for thermoforming.

Highly cross-linked thermoset plastics are very limited in their thermoformability. Only a few special postforming grades can be formed into angles or channels by heating and application of powerful mechanical forces. The postformable phenol- and urea-formaldehyde laminates fall into this category. Their thermoformability is restricted to linear bends at relatively large radii.

Specific heat

There are many thermal properties to consider, especially when sheets of different materials are used. Every substance has its own characteristic specific heat, that is, the amount of heat required to increase its temperature by 1°. Water serves again as the standard, receiving the number 1, to which all other materials are compared. For instance, polystyrene has a specific heat of 0.32, which means that 0.32 Btu is required to heat 1 lb of it by 1°F. In Table 2.1 (page 16) the values of a few plastics are listed in comparison with other common materials. With the exception of water, most materials have a lower specific heat than plastics. A more detailed listing can be found in Appendix A, which presents properties of a wider range of thermoformable plastics.

To use these values, one must realize that they represent a weight relationship. Because substitutions are mostly done without changing the gauge thickness, the variations in specific gravity must be taken into account. The specific heat, which can be accurately determined by subjecting the material to a differential scanning calorimeter test, increases slightly with rising temperature and jumps markedly when crossing the glass transition temperature. Therefore, listed values must cautiously be applied for calculations spreading over a wide temperature range. Published data can show great variations whether measurements were taken at room temperature or at a temperature range related more to thermoforming.

Thermal conductivity

There are also differences among plastics in regard to heat conductivity. These values are listed as Btu·ft/sq ft·hr·°F—the number of Btu, which are conducted through each square foot in 1 hour if the temperature difference is 1°F and the thickness of the piece is 1 foot. The contrast in values between plastics and metals is apparent. Especially when thick sheets are being formed, the low conductivity of plastics will limit the heat energy transfer. With excessive heat inputs the plastic surface may blister or start to scorch, even though the center region has not yet reached its softening stage. That is why sandwich heaters are recommended and why a brief delay between the heat and forming cycle may be beneficial for letting the heat soak into the center of the sheet. According to newer observations heating times can be reduced for heavy-gauge sheets when shorter wave radiation (visible light), which penetrates deeper into the sheet, is being employed.

Thermal expansion

The linear coefficient of thermal expansion is expressed in in./in.·°F. Because the plastic sheet expands more in the center, a wrinkling of the sheet is observed during the first part of the heating cycle. These wrinkles later disappear when the sheet contracts, due to the reversion of the drawdown imposed during the extrusion process. For practical purposes, the expansion of the plastic during heating can be disregarded. This is not true for the opposite, thermal contraction during cooling, which is manifested in part shrinkage. That is the reason why molds must be fabricated with oversized dimensions. The numerical values of the coefficients for thermal expansion and contraction are identical. Problems may arise when the formed parts must fall within very narrow dimensional tolerances, e.g., when they must be combined with machined metal parts or when closures are formed that must match the container size. Another problem may arise when shrinkage occurs on male molds and the formed parts become difficult to remove. Plastics with a high coefficient of thermal expansion in the range of 7 to 10 $\times$ 10^{-5} in./in.·°F are the cellulosics, polyolefins, and plasticized polyvinyl chloride; most other plastics lie in the range of 3 to 5 $\times$ 10^{-5} in./in.·°F.

Heat of fusion

An additional complication in regard to the above described thermal properties is introduced when crystalline or—more accurately expressed—semicrystalline plastics are being considered. Because most crystalline regions are larger than the wavelength of visible light, crystalline materials can easily be recognized by their translucent or white opaque appearance as long as the material is unfilled and unpigmented. As a matter of fact, the great differences in the degree of crystallinity can easily be noticed on the polyethylenes: the almost clear low-density polyethylene film has approximately 30% crystallinity and the white opaque high-density polyethylene molded parts have 90% crystallinity. When heated, these materials absorb not only specific heat as described above but also the heat of fusion for the content of crystalline material in the plastic, once the melting point is being exceeded. Thus, they require an extended heating cycle. Unlike water, the crystallization and fusion of plastics occur within a wide temperature range. With increasing temperature, fusion starts slowly and accelerates until the crystalline melting point is attained.

Several changes take place near the crystalline melting point. If unpigmented, the plastic becomes optically clear. Simultaneously, it is transformed into a limp material. After passing through the wrinkling stage, the sheet expands markedly and begins to sag.

Most of the difficulties arise after the forming stage is completed because a longer time period is required to disperse the greater amount of heat energy. The formed part tends to shrink more tightly to the mold and thus is harder to remove. Though crystallization commences rapidly, shrinkage to final dimensions will take days, because the rate of recrystallization decreases as the temperature drops. Rigid plastics will stop crystallizing once cooled below a certain temperature. Proper design of the mold-cooling system can ensure uniform cooling, thereby forestalling warpage in formed parts.

Thermal diffusivity

The use of the material constant, thermal diffusivity, would be ideal for establishing cooling times for thermoformed parts because the time required for cooling the heated and formed plastic sheet is proportional to the second power of the material thickness and inversely proportional to its thermal diffusivity. Furthermore, thermal diffusivity is clearly defined by its relationship to other establishable constants:

$$\text{Thermal diffusivity} \frac{\text{in.}^2}{\text{sec}} = \frac{\text{Thermal conductivity} \frac{\text{Btu} \cdot \text{in.}}{\text{sq in.} \cdot \text{sec} \cdot {}^\circ\text{F}}}{\text{Density} \frac{\text{lb}}{\text{cu in.}} \times \text{Specific heat} \frac{\text{Btu}}{\text{lb} \cdot {}^\circ\text{F}}}$$

However, a problem arises when one considers that all three material constants are not constant over the whole temperature range encountered in thermoforming. In addition, the latent heat of fusion becomes absorbed into the thermal diffusivity too. Therefore, published values for thermal diffusivity vary widely, depending on the temperature limits selected for their determination.

For this reason, industry still relies more on practically established cooling times rather than on mathematically calculated values. Knowledge of the mathematical relationship is still useful when dealing with just one or two changes among the many parameters. Examples are an increase in sheet thickness, higher mold temperature, substituting a material of lower density, etc.

Thermal stability

The thermal stability of plastics is of great importance for injection molding. To a certain degree it is also significant for thermoforming. Although forming temperatures are much lower than thermal decomposition temperatures, the plastics can be harmed if heat is applied too forcibly, due to its localization at the surface. As indicated previously (page 27), conventional methods for determining the temperature of plastics (thermocouples or thermistors) are not suitable for checking surface temperatures.

Water absorption

Thermoformable plastics vary greatly in their capacity to absorb water. Some plastics, such as the polyolefins, absorb almost no water, whereas cellulosics and nylons may absorb it assiduously. Although no problems based on water absorption may appear under normal conditions—on occasion unpredictably—it can interrupt production. Therefore, it is important to understand this phenomenon. Freshly extruded film and sheet tend to be completely dry. Even if a sheet were submerged in water for a short period of time, the material would still be practically bone dry. Water absorbed just at the surface will rapidly dry off during heating of the sheet. The surface-absorbed water will only slowly permeate (migrate) into the center part of the sheet. Although listed water absorption data are usually determined after 24 hours of submersion at room temperature, saturation may occur only after weeks or months.

Depending on the relative humidity of the ambient air, the water content of a sheet—if stored detached freely—will vary significantly. Fortunately, several months of high humidity are required for moisture to penetrate tightly stacked sheets or rolled film. Problems may arise only with the first outer sheets used after a weekend or other lengthy production stoppage, especially in spring when warm moist air enters the cool warehouse.

The slow rate of water permeation represents also the cause of difficulties during thermoforming. Although the moisture in thin films or on the surface of sheets will rapidly escape during heating, the absorbed water in the center of heavier sheets stays trapped and vaporizes inside the sheet, forming an opaque haze or bubbles of various sizes. The result is that the formed parts will appear to be hazy or foamy and display a rough surface or pock marks.

Varying with the thickness of the sheet, a drying time of one to several hours in an air convection oven at a temperature below the heat distortion point will be required. Also, sheets must be dried, individually supported, and never stacked. Unfortunately, for many processors this is not worth the effort, and such sheets are just discarded. Probably it would be more practical to store such rolls (wrapped in paper only) for a period twice as long as they were exposed to high humidity (months or years) in a boiler room or any other year-round dry, heated room. If only a few pieces are needed and no ovens are available, drying can be accomplished in the thermoforming equipment by the use of several repeated very brief heating cycles. It is always advisable to keep rolls and sheet stacks of moisture-absorbing plastics tightly wrapped—usually in two layers of polyethylene film—whenever not used.

Chemical effects, which means breakdown of the polymer molecule due to moisture content, are not to be expected during thermoforming. But because webs and edge trim are mostly utilized during reextrusion of the material, thorough drying is necessary for all the plastics subject to hydrolysis (polycarbonates, polyesters, urethanes, etc.) before reprocessing.

Water absorption data for plastics are listed also in Appendix A. But its magnitude is not proportional to the trouble it can provoke. In materials listed with values greater than 1%, the water is acting as a plasticizer and can be seen essential for warding off brittleness (nylons and cellulosics). A polycarbonate sheet having just 0.1% moisture content will blister during thermoforming, whereas a 5 times higher moisture content in cellulosics or nylons is readily tolerable. Materials with values less than 0.03% should not be expected to cause trouble. The detrimental effect of moisture on materials depends very much on the rate of absorption and the rate of hydrolysis at the processing temperature.

Orientation and crystallization

Orientation and crystallization of polymeric materials bring about peculiar arrangements of polymeric chain segments in the otherwise amorphous base material. Their great effect on material properties makes it essential to know and control them. As is often the case, generalized statements lead to many misconceptions. Therefore, this type of phenomenon should be discussed in its entirety.

When cooled, a molten polymer will first solidify with a random arrangement of polymer chain links. This material is called an amorphous material and is characterized by the fact that its properties are isotropic, the same in length, width, height, or any other direction.

All synthetic fibers that consist of polymers exhibit a high degree of orientation in the longitudinal direction only. One discovered that when orienting a solid fiber lengthwise by stretching it 30 to 300%, its tenacity can be increased manyfold. In this state many of the polymer chains become arranged in an orderly fashion but should not be regarded as truly crystalline because the order extends only in one direction. True crystals are ordered in all three directions. The stretching orientation is performed at a temperature that is lower than the melt temperature but higher than the glass transition temperature. This stretching requires considerable force. The chain links become not only oriented but also arranged closer together, causing other property changes (e.g., specific gravity, solubility in solvents, etc.).

On the other hand, hardly any force must be exerted to form or stretch a molten polymer. The coiled, randomly arranged polymer chains will just slide by each other with minimal aligning or orientation of the chain segments. The applied force is just required to overcome viscosity. The end properties of the formed plastic are, therefore, only minimally affected.

To illustrate these differences in orientation, two examples each are given for both polystyrene and polymethyl methacrylate.

Polystyrene extruded as a sheet will have an orientation due to its flow through the die and the gentle pull of the takeoff rolls. This can be proven by rotating a piece of the sheet between two crossed, light-polarizing films. Orientation can also be demonstrated more quantitatively by supporting a square of the sheet with known dimensions between thin plates coated with polytetrafluoroethylene or covered with talc and exposing it for several minutes to a temperature approximately 50°F below the melt temperature. The amount of shrinkage will be proportional to the amount of orientation, approximately 10% in the extrusion direction and less than half of that in the cross direction. Excessive orientation may pose problems in thermoforming if the sheet is not clamped tightly, or it may cause excessive mold shrinkage in the orientation direction, resulting in distorted parts.

If an extruded polystyrene sheet is given a biaxial orientation by stretching it forcibly at distinctly lower temperatures than its glass transition temperature (220°F), the resulting biaxially oriented polystyrene (OPS) sheet will have drastically altered properties. The sheet has become puncture and tear resistant and will not fracture when folded and creased; its surface has become harder and more scratch resistant. Both kinds of polystyrene sheet can be thermoformed. The latter in thin gauges has found wide application in packaging.

Cast polymethyl methacrylate sheet is a truly amorphous, isotropic material. If it is heated properly and thermoformed, the material of the shaped part will have practically the same properties as the original sheet. Still, that the material has been deformed during the thermoforming process can easily be proven by placing the formed part back into the oven, whereupon it will return to the original, flat sheet configuration. This phenomenon is generally called *plastic memory*. This experiment will not necessarily work with any other plastic. Both forming (stretching) and retraction take place under no or very low forces.

On the other hand, if forming of polymethyl methacrylate (PMMA) takes place at a much lower temperature (below the glass transition temperature), great mechanical forces must be employed to orient the polymer chain links in the direction of pull. The resulting material (when oriented in two directions) will be very tough and is therefore used for bullet-proof laminates. This kind of orientation involves high internal stresses. When improperly balanced, a PMMA part that comes into contact with certain chemicals can become crazed and cracked. For this reason, part surfaces and edges are usually protected. Generally, the thermoforming of polymethyl methacrylate parts should always be done at sufficiently high temperatures to prevent the possibility of crack formation during later use. This is also recommended for many other materials, e.g., polyethylene terephthalate.

The biaxially oriented film materials that have reached high-volume applications for packaging are polyethylene terephthalate, polypropylene, polystyrene, and polyamide.

Similar drastic changes in properties are experienced in several forming processes that compete with thermoforming, such as solid-phase forming and cold forming (see Chapter Twelve).

On the other hand, with some polymers, crystallization will take place spontaneously—without applying external forces—but never instantaneously. For crystallization to occur, the material's temperature must be lower than the melting point but not too low, since the chain segments must remain mobile enough to arrange themselves in a three-dimensional orderly fashion. Due to restraining forces of the angular chain link interconnections, polymer materials cannot crystallize completely. Because the chain links within the crystalline region are more closely spaced than in the surrounding amorphous regions, semicrystalline polymers always consist of a mixture of two different identities. They are (with some exceptions) opaque and become transparent only when heated close to or above their melting point. The higher the crystalline content in a polymer, the higher will be the specific gravity and the modulus of elasticity (rigidity), but brittleness can increase too. Crystallization can sometimes be prevented or reduced by rapid cooling (quenching). If chain segment alignment has been accomplished by orientation, alignment to orderly crystallites is usually excluded. Caution: crystal polystyrene is not at all crystalline. It was so termed because parts molded from it have a "crystal" clear appearance. Semicrystalline polystyrene, having a (syndiotactic) stereospecific arrangement of the styrene monomer units, is just now, under the trade name Questra by Dow Plastics, being developed for special engineering applications. Its heat deflection temperature of 210°F (versus only 170 to 200°F for amorphous polystyrene) is comparable with that of engineering thermoplastics.

Light can be shed on these complexities by considering the behavior of several currently available thermoplastic polyesters. First, a low molecular weight poly(ethylene glycol terephthalic acid) ester was produced. This material was unsuitable for molding useful parts because this low molecular weight material became transformed within a few days into a highly crystalline substance that was opaque and very fragile.

When heated above its crystalline melting point, it converts into a viscous liquid. However, it matured into very important materials when its tendency to crystallize could be suppressed by orientation. The widely known polyester textile fibers are gained by highly orienting unidirectionally the melt extruded (spun) fibers. The very tear-resistant, dimensionally stable, and nearly clear polyester films represent extruded films that are biaxially oriented before crystallization would occur. Both are products of the same polyethylene terephthalate homopolymer resin.

Stretch-blown polyester plastic bottles, commonly known as plastic containers for carbonated soft drinks, use the same chemical composition but of a higher molecular weight. A very simple experiment can clearly demonstrate the interrelationship between those different structural phases that are exhibited by chemically identical polyester segments. As seen in Figure 8.2, the injection-molded preform (at left) represents a clear part that consists of randomly arranged polymer segments. Its mechanical properties are isotropic because no orientation has yet taken place. It is clear because crystallites have not been able to form due to rapid cooling of the melt. After quick reheating the body of the preform is stretched to form the bottle (center) but the neck remains unchanged. When such a bottle is placed into an oven, the temperature of which is held at below the crystalline melting point (490°F) but above the glass transition point (170°F) for some time, the polymer segments become somewhat mobile again. First, the highly stretched areas will retract, moving the material back to resemble the shape of the preform. The polymer segments in the unstretched neck of the bottle will slowly turn white (at right) due to crystal formation. At extreme time-temperature exposures all the material will crystallize into a completely white appearing mass.

By exchanging the ethylene glycol with butylene glycol the chain segments are altered. Materials of this composition are semicrystalline

Figure 8.2. Stretch–blow molded polyester bottle: (left) injection molded preform, (middle) blown bottle, and (right) much of the elastic stretch recovered and neck part crystallized on reheating.

and mainly used filled and glass fiber–reinforced as injection-molding compounds. A highly mineral-filled version, Enduran from GE Plastics, with its ceramic or marble look and feel is thermoformable and used for solid-surfacing applications in kitchen and bath, rivaling DuPont's Corian, the composition of which is not disclosed. The processing of these sheet materials at 1/2 in. thickness represents the extreme in production time requirements. At 350°F the heat-up time can stretch up to 1 hour and the following cooling time half an hour.

Later on, many more thermoplastic polyester formulations that have found wide applications both for injection molding and thermoforming with great emphasis on the packaging sector were introduced. Various schemes have been found to obtain the wanted properties of the final product by regulating or completely eliminating the crystallization process. This has been accomplished by disturbing the regularity in the polymer chain with the introduction of differently shaped monomers or by compounding in micronized filler particles. The ethylene glycol component (E in PET) may be substituted by cyclohexanedimethanol or cyclohexylene glycol (C in PCT and also G in the copolymer PETG) and the terephthalic acid by either isophthalic acid or 2,6-naphthalene dicarboxylic acid (N in PEN for polyethylene naphthalate ester). It is important to remember that, with the increasing molecular fraction of ring segments or ring size in the polymer structure, the use temperature becomes elevated. Thus, the melting points of the various polyesters increase from polybutylene terephthalate (PBT) 437°F, over PET 482°F, PEN 523°F, to finally PCT with 545°F. Other desirable properties are often also improved. The gas barrier properties of PEN are approximately 5 times better than those of PET.

Due to the great variability of possible combinations caused also by the multiplicity of copolymer formulations, the chemical identification of commercial products becomes confusing. The crystallizable polyethylene glycol terephthalate (CPET) contains a nucleating agent that speeds up crystallization. However, due to a remarkable increase in molecular weight the crystallinity content can be restricted to approximately 30%. This results in a product combining some optimal properties:

(1) Higher rigidity and better temperature resistance
(2) Good low-temperature toughness
(3) Good barrier properties

These attributes have made CPET an ideal material for opaque, ovenable food-packaging trays. They are suitable for cooking of food in either a microwave or a conventional oven. When consulting Table 2.2

(page 28), one may notice that this polyester is processed under conditions resembling those used for thermoset materials, with the mold being heated up to 350°F. Although no chemical reaction takes place, the high mold temperature is needed to carry out the crystallization process within a short time span.

Due to the high-temperature rigidity of the crystallized polyester it is impractical to thermoform an already crystallized sheet or film. Thermoforming the noncrystallized material into a desired shape by conventional methods and then attempting to crystallize the formed part by heating would grossly distort its shape and would require a long processing time. Therefore, the forming process is combined with the crystallization step. Figure 8.3 illustrates that the fastest crystallization occurs between 300 and 360°F. Crystallization will start when the sheet is heated in the oven but must be continued on a heated mold (250 to 350°F) for 2 to 6 seconds. The progress of crystallization can be followed by observing the onset and intensity of the opaqueness of a clear (unpigmented) sheet material. Because the properties of the formed part are very dependent on the extent of crystallinity, the establishment of its value should be a part of quality control. The percent crystallinity can easily be established by the determination of the sheet's density. This

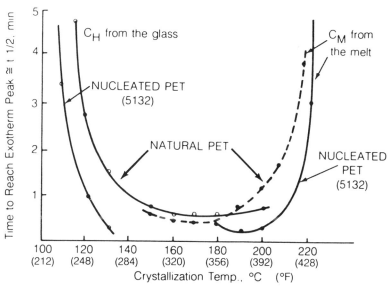

Figure 8.3. Crystallization half-time of natural and nucleated PET. The inverse crystallization rates for PET are shown on the left-hand side as they are observed when heating a sheet. The crystallization rates observed when cooling a melt of this polymer are shown on the right-hand side (courtesy of Eastman Chemical Company, Kingsport, TN 37662).

nearly linear relationship indicates at 77°F a density of 1.334 g/cm³, a crystallinity of 15%, and at 1.367 g/cm³ one at 45%. The lowest range of 15% might be selected if emphasis is placed on low-temperature toughness and the highest of 30% if high-temperature dimensional stability is required.

Figure 8.4 shows a two-step mold where the first row consists of a heated set of female molds for the forming and crystallization step, whereas the second row of cooled molds receives the previously formed parts. The process is completed by preblanking, cooling, and final trimming of the finished parts within the same time span. A formed part is shown resting on the top mold platen. A computer-regulated drive motor ensures precise advancing of the web. Four-column guides are recommended to absorb the unbalanced forces in those combination tools.

Another class of ethylene terephthalate (co)polyesters are generally marked as APET or amorphous polyesters, which, however, embrace also many of the amorphous copolymers, such as PCTA, PETG, or PCTG. All of these film materials are clear and not as rigid and strong as the original biaxially oriented polyester film. But some of them excel in regard to their upper use temperature. Those polyesters that cannot crystallize remain clear, and orientation is not required. These amorphous thermoplastic materials are widely accepted in the production of

Figure 8.4. Four cavity tool for crystallizable polyester trays (courtesy of Gabler Maschinenbau GmbH, D-23512 Luebeck, Germany).

thermoformed packaging, primarily competing with the more expensive polycarbonates.

Both orientation and crystallization disrupt the loose, relaxed arrangement of polymer chain segments in an amorphous material. If both occur in short chain length polymers, they lose elasticity and elongation, becoming quite brittle or fragile and, therefore, useless. In very high molecular weight polymers, the individual chains will sufficiently span several of these ordered (crystalline) regions, so that amorphous polymer properties (high elongation, ductility, etc.) are adequately retained, making them very useful products. The old concept of chain folding does not agree with the observed crystallization process or with the enhanced material properties (G. Gruenwald, Proposed structures and models for polymers crystallized in bulk. *Journal of Polymer Science,* 61:381–402, 1962).

As in many other cases, contrasting properties are achievable with oriented and crystalline materials. For a great number of packaging applications, high shrinkage (significant dimensional change) is needed to hold objects tightly or to expel surrounding air. Heat-shrink film and tubing fall into this group. In other uses, such as graphic film bases, the slightest shrinkage or distortion becomes intolerable. Because for both exactly the same highly oriented material is used, the material for the latter application must be heat-set by holding it taut for a few seconds at a temperature markedly higher than it will ever be exposed to in use.

To stabilize highly oriented plastics, one of two conditions must be fulfilled:

(1) The glass transition temperature must be higher than the temperature of use
(2) The orientation must be fixed by the formation of quasi-crystalline regions

Many of the useful oriented film materials belong in the latter group, e.g., polyolefins, polyvinyl fluoride, and the vinylidene chloride copolymers.

Although most plastics can be separated into the two groups of amorphous and semicrystalline plastics, there are several other exceptions, among them polyvinyl chloride and polycarbonate. The latter is, when thermally processed, always amorphous but can also crystallize when treated with solvents.

These examples underscore the importance of knowing which property changes may occur whenever orientation or crystallization processes are involved in thermoforming. Some of these changes can appear unexpectedly and are usually not desired. To avoid unpleasant failures in thermoformed parts, the need for extensive implementation of long-term environmental testing cannot be overemphasized.

Manufacture of starting materials

There are two reasons for the thermoformer to become familiar with the various processes employed by manufacturers of film and sheet materials. First, different processes may require variations in formulation, such as stabilizers and lubricants, and they usually demand polymers of different molecular weight and melt viscosity. Second, the suitability for thermoforming of materials from different manufacturers may vary. Changes in thermoforming processing must invariably be made when using film or sheet of different origin, even though the basic plastic material remains the same. These differences are primarily caused by variations in melt viscosity and frozen-in stresses. Stresses released during the thermoforming process may also have an effect on shrinkage or warpage of formed parts.

Most available sheet materials are produced by the screw extrusion process, which employs medium to high molecular weight polymers that should be subject to minimal heat stress. Because pull-down is required to maintain a constant thickness, the produced sheets contain a certain degree of stretch (approximately 10%). This will become noticeable when the sheet is being heated prior to thermoforming. The sheet will tend to shrink in the machine direction (approximately 5 to 15%) but possibly expand somewhat (0 to 5%) and thus sag in the cross direction. The degree of stretch can easily be determined by placing a square piece of it on a silicone or polytetrafluoroethylene-coated metal sheet for 15 minutes in an oven at a temperature approaching the forming temperature.

Thinner films are usually produced by the chill roll casting process in which the polymer is heated to a higher temperature, but less strain is incorporated in the film due to the short distance between die and chill roll.

Polyethylene film and some olefin copolymers are produced in large volume by the blown film extrusion process. The melt is extruded through an annular die and blown up into a very large-diameter bubble. The film is then collapsed, folded or slit, and wound up in rolls. Because the polymer is stretched in the longitudinal and transverse direction, the obtained film has good mechanical properties. These products, however, differ considerably among various suppliers.

Vinyl sheets, a few polyethylene, high acrylonitrile, and acrylonitrile-butadiene-styrene polymer sheets have also been produced by a calendering process. This process requires a considerably higher capital investment and is, therefore, restricted to very high-volume usages. Usually, even higher molecular weight resins are used. The compounded material is first homogenized in high-intensity mixers, planetary gear extruders, or on two-roll mills and then the sheet formed in the lip between hot rolls of a four-roll stack. Further smoothing occurs while passing through two more closely held gaps. These sheets are free of frozen-in stresses and have very good surface smoothness. Better gauge control, the availability of sheets with wider width, and the greater productivity results in lower cost products. They may also be available with embossed patterns on one side.

Biaxially stretched films (e.g., polystyrene) behave differently. Upon heating, they exert a high-shrink force and, therefore, must be clamped or held very tightly to the heating surface, as is the case in the trapped sheet pressure-forming process (see page 168).

Sheets with excellent optical properties are obtained either by a casting process (e.g., polymethyl methacrylate sheets, which are still cell cast or continuously cast between two endless steel belts) or by laminating and/or press polishing of otherwise manufactured sheets. This extra processing step, however, adds appreciably to the cost of these materials. Because surface gloss cannot be imparted by the thermoforming process, these higher cost materials must be used if high gloss becomes a requirement for the formed part.

Coextrusions and laminates

Coextruded and laminated sheets have gained favor for thermoforming processes because for many applications no other practical molding process can duplicate those properties so easily obtained by thermoforming.

By the term *coextrusion* is meant the formation of a sheet product by simultaneously employing two or more extruders. It is not necessary to feed all the resins through one die. The layers may also be combined from two or more dies externally as long as the materials are hot enough to bond to one another.

The term *laminate* should always be employed when either two or more previously extruded or calendered sheets are combined or bonded, or when an already extruded sheet or material layer is coated with an additional layer of a polymer resin. Originally, laminates were obtained by pressing several layers between chrome-plated steel sheets. Now, they are normally made by only brief line-contact pressure between rotating heated rolls.

It is impossible to list all material combinations; therefore, emphasis will be placed on rationale of employing coextruded sheets.

Coextruded products and laminates are used when:

(1) Parts are sought after which have different colors on the inner and the outer surface, and when just a plain coat of paint is not satisfactory.

(2) Parts are subject to ultraviolet radiation on the outside, but a lower cost material suffices to provide mechanical strength. Three outstanding examples in this area are: (a) the acrylic multipolymer film (see Korad in Appendix C) laminated to acrylonitrile-butadiene-styrene (ABS) or other sheets; (b) coextruded sheets consisting of acrylonitrile-styrene-acrylate on the outside over (high impact) ABS; and (c) polyvinyl fluoride film (see Tedlar in Appendix C) bonded to ABS or other thermoplastic sheets.

(3) Low-cost regrind or scrap materials (of undesirable color) are available, but parts with a high-quality appearance are demanded. Appearance and, to a great extent, mechanical properties, too, are determined by the quality of the outside layers of a sheet. Inferior-quality center layers may have negligible detrimental effects, as long as good bonds between the layers can be ascertained.

(4) No material that could provide all the properties required for the particular application is available. In food packaging, where low-oxygen permeability, low-moisture permeability, and heat sealability must be provided, coextruded films have gained wide usage. These coextruded films will be further described in the paragraphs on barrier materials.

The one problem arising with these combinations of materials is that the trim materials have a limited use after regrinding (see example 3 above). In the case of an in-line thermoforming process this can be lessened somewhat by using an uncolored or waste resin for the sheet edges which will just pass through the transport chains and will remain always outside the formed part. From the trim skeleton these edges can then be separated and reused as regrind.

Mechanical properties

In selecting materials, mechanical properties are probably the most basic. It is not peculiar to plastics that different materials are recommended for the same application, just as houses can be built of wood or bricks. For instance, polyvinyl chloride and polyester films are advertised for the same packaging applications.

Fulfilling the rigidity and impact requirements is indispensable for any part. The widely promoted tensile strength properties are actually of little value to the designer. The modulus of elasticity under flexure is the best expression of relative rigidity and probably the most valuable data in connection with some of the thermal properties and impact properties, frequently listed in materials property tables.

It should be emphasized that for structural parts expected to serve for several years, long-term performance data should not be overlooked. For certain materials, graphs can be found which relate to and sometimes compare long-term performance. This kind of behavior cannot be expressed in simple numerical values, as the other above-mentioned properties. Consequently, even for the expert, it is difficult to apply valid data for comparisons. The three primary factors should be briefly cited:

(1) *Creep,* the slow deformation occurring when plastic parts are subjected to a constant force at a certain temperature. This can change the shape of the part in time.

(2) *Fatigue,* sudden failure in a part subjected to long periods of cyclic loading or vibrations.

(3) *Aging,* the slow degradation of properties due to the influence of environmental factors (temperature, chemicals, radiation) acting on the part material. Exposure to strange environments combined with external loading may cause sudden failure, which is more related to environmental stress cracking.

A designer who is aware of these three effects that cause the majority of material failures can assess the intended application and seek a comparable application proven over time. If one cannot be found, he may devise a test to assure himself that the new design is practicable.

Material economics

Due to the wide range of thermoforming processes no attempts have been made in this book to describe examples for cost estimating, because it is highly unlikely that assumed parameters will be comparable with the condition encountered by the reader.

Only after hesitation were prices of the various generic materials included in the property charts. Despite low inflation and low interest rates in the early 1990s, further price fluctuations may still occur. Current prices for generic materials may span a rather wide range. For the price tabulations, quotations obtained for general purpose extrusion grades, in natural color and truckload quantities, were considered. Gauge disparities in films hamper the comparison among different materials and applications.

To facilitate calculations, the more important cost of the material per cubic inch has also been included. This number was obtained by multiplying the price per pound by the specific gravity and a factor of 0.0361 according to:

$$\frac{\$}{\text{cu in.}} = \frac{\$}{\text{lb}} \times \text{Specific gravity} \times 0.0361$$

This formula is based on the fact that when materials are replaced, dimensional variations are, at first, not a factor. However, because the replacement material may result in differing properties, additional aspects can affect the actual part cost. Two of these should be elaborated:

(1) Rigidity: Most parts used for any structural application (including packaging) are designed for certain minimum rigidity values at the highest temperature of use and/or a minimum resistance to impact. Therefore, if the replacement material is superior in these respects, reductions in wall thickness should be considered concurrently.

(2) Processing: Shear resistance and heat stability (among other properties) will greatly influence extruder output and the reprocessability of trim regrind. The latter is especially important for many thermoforming processes. Again, if the replacement material ensures either higher outputs or higher yields, savings in processing could be realized.

Some important formulas should be specified, because the designer must always keep in mind the relationship between sheet size and weight. The specific gravity of the material, which is contained in any materials brochure or handbook (see Appendix A), must be ascertained to determine the sheet weight or the weight per square foot.

$$\text{Sheet weight: } \frac{\text{lb}}{\text{sheet}} = \text{Length, in.} \times \text{Width, in.} \times \text{Thickness, in.}$$

$$\times \text{ Specific gravity } \frac{\text{g}}{\text{cm}^3} \times 0.0361$$

$$\text{Weight per square foot: } \frac{\text{lb}}{\text{sq ft}} = \text{Specific gravity } \frac{\text{g}}{\text{cm}^3}$$

$$\times \text{ Thickness, in.} \times 5.20$$

Example: A pigmented acrylonitrile-butadiene-styrene copolymer sheet 0.125" thick is listed as having a specific gravity of 1.07 g/cm^3. A sheet 2 ft. by 3 ft. will weigh:

$$24 \text{ in.} \times 36 \text{ in.} \times 0.125 \text{ in.} \times 1.07 \frac{\text{g}}{\text{cm}^3} \times 0.0361 = 4.17 \text{ lbs}$$

One square foot of this sheet weighs:

$$1.07 \frac{\text{g}}{\text{cm}^3} \times 0.125 \text{ in.} \times 5.20 = 0.695 \frac{\text{lb}}{\text{sq ft}}$$

To determine the yield in square feet per pound the following equation should be used:

$$\frac{\text{Square feet}}{\text{Pound}} = \frac{0.1923}{\text{Specific gravity } \frac{g}{cm^3} \times \text{Thickness, in.}}$$

$$= \frac{0.1923}{1.07 \frac{g}{cm^3} \times 0.125 \text{ in.}} = 1.438 \frac{\text{sq ft}}{\text{lb}}$$

If the material is used in roll form, the following formula is helpful to determine the remaining length of film in feet:

$$\text{Length of film, ft} = 65.45 \frac{(\text{Outer dia, in.})^2 - (\text{Inner dia, in.})^2}{\text{Film thickness, mils}}$$

Example: Polyester film of 7 mil thickness with outer diameter of 4 1/4″ on a 3″ roll (inner diameter):

$$65.45 \frac{4.25^2 - 3.1875^2}{7} = 65.45 \frac{18.06 - 10.16}{7} = 73.9 \text{ ft}$$

Regrind utilization

Thermoforming represents the only plastics processing method where commonly 50% of the starting material might end up in a nonintended shape, namely as a web or skeleton, after the produced parts have been stamped out, or in edge trim and cutouts. To such losses, scrap parts during start-up and other end losses must be added. This holds true whether the parts are made on a roll-fed, thin-film machine or heavy-sheet rotary thermoformer. This necessitates, unless just a few pieces are to be made, that provisions are taken to utilize this large amount of trim. If the thermoformer is in-line with a film extruder, all trim material can easily be returned, pelletized, and fed into the extruder by maintaining a steady ratio of virgin and reground material. Such closed-loop reprocessing can be performed endlessly.

At installations where thermoforming takes place away from the extruder, being in another building or another geographic location, the trim material must be collected and shipped back to the extruder. The main problems arise due to the need of preventing any contamination of the plastic. This is difficult to accomplish due to static electricity attracting everything the plastic comes in contact with. The other problem is that these two processes may become out of sync, thus making it difficult to maintain a set virgin-regrind ratio. If this ratio stays high in favor of regrind, contamination and degraded polymer may accumulate over time and may make it difficult to maintain high-quality specifications. Under those conditions it might become advantageous to seek applications where mere regrind material will suffice. After cascading down in properties a few steps, eventually some of the resulting material will end up being discarded or fed into other recycling schemes; see the end of the next chapter.

As stated elsewhere, the thermoforming conditions are seldom so severe that molecular breakdown might occur. This usually takes place during extrusion where the material is exposed to high pressure, shear and high temperatures, and the danger of introducing moisture with an improperly dried regrind material. Polyolefin resins generally degrade due to thermal oxidation and result in a material impossible to process due to excessive sagging of the sheet. Polyvinyl chloride and other chlorine-containing plastics suffer from hydrogen chloride formation. The reduction in molecular weight and with it the deterioration in properties becomes first noticeable in the drop of impact strength or in yellowing of the material. Only severe degradation will also show up in loss of modulus of elasticity and yield strength.

NINE
Thermoforming Materials' Chemical Descriptions

BEFORE SOME KEY properties of the most important plastic materials are described, a few remarks that may apply to any of them must be made.

Appendix A lists most plastics used for thermoforming processes grouped by chemical name to avoid favoring any one brand. Because in thermoforming applications commodity resins are often used, this tabulation appears logical even though in a few cases only one commercial product in a group is available. On the other hand, some important products are generally not known by their chemical composition, and some are still being sold without any indication of their composition. To help the reader in these cases, another list has been added to the Appendices. This supplemental list, Appendix C, identifies proprietary plastic products in alphabetical order using proprietary trade names—some of which are registered names—along with the manufacturer's name, chemical composition, and application (thermoforming, packaging, or special machinery parts).

For the sake of consistency and avoidance of misunderstandings, all chemical names are spelled out. Occasionally, the better known abbreviations are added in brackets.

Numerical values are taken, where possible, from manufacturers' properties charts and from handbooks. Unavoidably, not all these values will be representative of a specific material within a group. Those plastics—mainly copolymers and polymer blends or alloys—which exhibit so wide a diversity in properties that examples of them have little similarity, are difficult to tabulate in a consistent way. Therefore, either the average values, the most representative values, or a range of values has been listed for them. This class of materials is also the one that has grown rapidly during recent years and is still expected to continue to grow in the near future.

In any case, the reader should make final judgment with property values obtained from the manufacturer whose product is intended to be used.

Because the opinions of observers differ in regard to the ranking in importance of the various materials used for thermoforming processes, these materials will be treated in related groups. Descriptions of them are brief and cover only general properties and behaviors. The final selection of a material from the vast number of commercially available materials is, of course, an individual decision that will include other criteria, as well as personal preferences.

Acrylics

Acrylics, which are more descriptively designated as polymethyl methacrylate, are well suited for thermoforming due to their high hot strength and wide processing temperature range of 290 to 350°F. No other material behaves similarly during thermoforming. Their excellent optical properties and clarity are augmented by outstanding outdoor stability and rigidity.

Cell cast acrylic sheets have the best optical properties and lowest shrinkage but are highest in cost.

Continuous cast sheets are produced by casting a viscous, partly polymerized monomer onto a metal belt. They are nearly as good in physical properties but may show slight optical distortions.

Extruded acrylic sheets are lower in cost but are not optically perfect and can have up to 15% orientation in the length direction.

Modifications of acrylics—many of proprietary composition—have mitigated their brittleness, reduced their cost, and improved many of their other properties.

Cellulosics

At one time the cellulosics were the preferred material for forming tough parts of acceptable clarity. But their high cost has tended to encourage the use of other clear materials. Although they are based on biomass raw materials, it remains questionable whether they might regain their position.

All cellulosics represent chemical reaction products of cellulose, which can be obtained from either wood or cotton linters. The name indicates which derivative it is, e.g., cellulose acetate, cellulose propionate, or cellulose acetate butyrate. However, because the cellulose macromolecule must be broken down during processing and then formulated with many other ingredients, mostly plasticizers, their properties may vary widely. It is, therefore, essential to specify the application when ordering cellulosic materials.

Some of their advantages are that they are easy to thermoform and never pose problems in trimming. They are especially appropriate for deep drawn parts. The cellulosics contribute strength and stiffness to formed parts, which can be transparent and also easily colored or decorated. Improved stabilization has made them suitable for outdoor applications, and they are available in food grades.

For thermoforming, cellulose-acetate-butyrate (CAB) has gained leadership, but cellulose-propionate has better clarity, toughness, and is easier to work with.

Polyolefins

Polyolefins comprise a great number of plastics, which have in common that they consist only of carbon and hydrogen atoms and do not contain any cyclic groups. They can be recognized by the feel of their surface, which resembles paraffin wax, their low molecular weight cousin. They are all resistant to water and aqueous solutions and may swell, but do not dissolve, in organic solvents unless heated.

Low-density polyethylene (LDPE), polymerized according to the high-pressure process, represents the earliest commercial product. This process resulted in highly branched polymers that are soft and translucent due to their low crystallinity. Variations and improvements in polymerization catalysts have yielded polymers with significant variations in densities, such as the high-density polyethylenes (HDPE), polymerized in a low-pressure solution polymerization process. The structure of these products consists of mainly linear chains that facilitate crystallization and are, therefore, white and opaque. Further developments have followed, making additional property alterations possible, such as the linear low-density polyethylenes (LLDPE) and ultralow-density polyethylenes (ULDPE). The newest metallocene catalysts, which are based on single-site, constrained geometry catalyst technology, allow the production of polyolefins with closely tailored properties due to those catalysts' capabilities of controlling molecular weight and copolymer structures within narrow ranges. But their higher prices must be balanced by their improved performance, and slight variations in their processing must be accepted. The preferred melt index values, which are reflective of the molecular weight, are for thermoforming polyethylene plastics only around 0.25.

Polypropylene and its copolymers have a low specific gravity of 0.90 g/cm^3 (and less), thus resulting in very high yields per weight. Their higher stiffness, based on their high degree of crystallinity, allows also the production of large parts even if they are destined to higher temperature uses. Therefore, polypropylene becomes frequently regarded as one of the engineering thermoplastic resins. Many impact-grade polypropylene copolymers are on the market for parts that must resist impact forces. Similar to polyethylene, improvements in polymerization techniques have led to low-modulus polypropylene grades, which exhibit a wider temperature range than presently used flexible polyvinyl chloride on one side and on the other extreme, to syndiotactic polypropylene that feature clarity and high-impact properties also at low temperatures. Regular polypropylene is quite brittle at low temperatures.

These advantages are augmented by their relatively low cost in comparison with other plastics. Still, it has taken a long time until polypropylene has obtained its remarkable market share for thermoformed parts. A relatively large amount of heat must be imparted to the sheet, due to the need of melting the crystalline fraction, to make it pliable enough for thermoforming. Similarly, a longer time is required for cooling. More problems are caused by its large thermal expansion during heating, which leads to excessive sagging of the sheet. This sagging can be offset by increasing the molecular weight of the resin, which, on the other hand, will reduce the output during the extrusion of the sheet. Other possibilities are the inclusion of fillers as reinforcing agents or the utilization of nucleating agents in small concentrations. That will lead to higher rates of crystallization and faster cooling due to the elevation of the endotherm temperature peak from 230 to 260°F. The smaller size of crystallites reduces the opacity of the product. The thermal expansion of the sheet can also be compensated by modifications of the thermoforming apparatus, e.g., providing preheaters or by outward diversion of the conveyor rails in the heating zone. The 3 times higher shrinkage versus polystyrene during cooling can cause problems in maintaining dimensional tolerances and in trimming accuracy. The trim-in-mold process can diminish that to a certain degree.

Polymethylpentene should be mentioned because it is the plastic of lowest density, 0.835 g/cm^3, and among all polyolefins the one with the highest crystalline melting point of 465°F (PP 335°F, HDPE 275°F, and LDPE only 240°F). Because density and the refractive index of both the crystalline and the amorphous polymethylpentene are identical, formed objects are generally transparent. Only its higher cost limits its widespread application.

Not even the copious styrene-derived copolymers can outnumber the many copolymers that evolved over time from polyolefins. Among the olefin monomers that have been used to alter some properties are propylene, butene, hexene, and octene. Any of these copolymers can be considered for thermoforming processes, as long as high molecular weight materials with a melt index of less than 2 are used. In regard to rigidity they may vary widely but are still on the softer side. These polyolefins are frequently used when impact resistance and toughness, rather than rigidity, are required.

The number of copolymers containing other than olefin monomers have been widely expanded by incorporating monomers that can generate polymeric materials of quite different properties. Some of these copolymers that broaden the application of polyolefins are ionomers (containing ionic carboxyl groups, important for coextrusions and for form-fill-seal food-packaging applications), polyallomers (used for integral hinge applications and food packaging), ethylene-vinylacetate (EVA) copolymers (used for packaging and as heat-sealing layers), and among still others the very important ethylene-vinyl alcohol copolymers (EVOH or EVAL). The latter outperform in regard to their barrier properties by far all the other barrier polymers. Due to their very high crystalline melting points (between 325 and 360°F) they are thermoformed either below their melting point by solid-phase pressure forming, yielding a high degree of orientation and mechanical strength, or at a much higher temperature according to the regular melt phase forming process. In this case deeper draw ratios can be achieved. Polyolefin plastomers (POP) combine good stiffness with low heat-seal temperatures, also benefitting packaging applications.

Polyolefins are not the easiest materials to thermoform. That is why many machine builders stress in their advertisements that certain kinds of their equipment are suitable, or especially designed, for forming polypropylene. The main reason for this difficulty is that the thermal conductivity of polypropylene is only one-third that of high-density polyethylene. In other chapters of this book the benefits and disadvantages of crystalline materials have been described extensively. Their forming temperatures are relatively high, which means that polyolefin plastics are much softer than most other thermoforming materials when the sheet touches the mold; see Figure 8.1 on page 113. Vacuum holes or slots in the molds must be smaller and the mold surfaces more perfect (but not smooth), to produce a part of good appearance.

Styrene polymers

Polystyrene is one of the oldest plastics and—based on its clarity and its rigidity—has been tagged crystal styrene. However, all commercial grades are amorphous, the syndiotactic crystalline form, having much higher thermal capabilities, is just now being introduced for engineering applications. Polystyrene's outstanding properties, apart from optical qualities, are a very high modulus of elasticity (stiffness) and excellent dielectric properties. Polystyrene parts can be recognized by their high ring: it sounds like metal when dropped onto a hard surface. Due to its fragility the name polystyrene fell into bad repute after other plastics with better impact properties became available. Extensive research yielded styrene copolymers and polystyrene alloys and blends. Yet still today, many resin manufacturers exploit the excellent properties of polystyrene in their products while avoiding mentioning its name in their advertising, sometimes even keeping it secret, as in some polyphenylene oxide alloy plastics.

Some of the styrene-containing plastics are:

(1) *Polystyrene* is a low-cost plastic mainly restricted to injection molded parts. Only biaxially oriented polystyrene and foamed polystyrene sheets have gained importance in thermoformed packaging.
(2) Some *styrene-butadiene plastics* (SB) are favored for clear packaging applications.
(3) (High) heat-resistant styrene copolymers and (high) impact styrene (HIPS) copolymers have gained great acceptance for thermoformed parts due to their ease of processing, reasonable cost, and good overall properties.
(4) *Styrene-methyl methacrylate* copolymers have improved outdoor and weather resistance.
(5) *Styrene-maleic anhydride* copolymers (SMA) dominate applications where higher heat stability, good molecular bond, and strength are required.
(6) *Styrene-acrylonitrile* copolymers' (SAN) largest application lies as a component in the manufacture of acrylonitrile-butadiene-styrene resins. Styrene-acrylonitrile's transparency, good chemical resistance, and heat resistance, has secured applications in many areas.
(7) The *acrylonitrile-butadiene-styrene* graft copolymers (ABS) exceed in importance all the above-mentioned styrene copolymers. They are available in so many different grades, making the selection of the right plastic complicated.

Equally impressive are the contributions of styrene polymers to the thermoforming process. Polystyrene has a low value for its specific heat, and it quickly regains rigidity when cooled after forming, allowing very short processing cycles. In addition, biaxially oriented polystyrene (OPS) and foam polystyrene sheet are major contributors to the thermoforming industry. One shortcoming, lack of ultraviolet resistance, has been overcome somewhat by improved stabilization and overcome completely by the extended use of coextruded cap sheets, consisting of acrylonitrile-styrene-acrylate copolymers (ASA).

Vinyl resins

Polyvinyl chloride is characterized by outstanding mechanical toughness, weather and chemical resistance, and inherent flame retardant properties. Because it retains its intermolecular structure despite the addition of low molecular weight plasticizers, it can be compounded to obtain any desired hardness or flexibility and still preserve its toughness and load-bearing capabilities. It is very well suited for thermoforming. Polyvinyl chloride itself and many of the other vinyl polymers containing halogen atoms, such as polyvinylidene chloride and fluoride, have gained importance in packaging and protective film applications, such as outdoor weather- and corrosion-resistant laminations.

The fluorinated ethylene-propylene copolymer, as well as the ethylene-chlorotrifluoroethylene copolymer, can be thermoformed and provide at a high cost excellent high-temperature properties and superb chemical and low-temperature resistance.

Engineering plastics

Although there are no set rules for deciding which plastics belong in this group, all members must have considerable toughness that is retained when exposed to elevated temperatures. For thermoforming volume usage they may not be as prominent as many of the commodity resins; however, they have been applied in a number of thermoformed parts where their higher cost can be justified.

The capability of these materials to withstand higher temperatures is mainly founded on the structure of their polymeric chains. The chain links of ordinary resins consist of small units, best characterized by the methylene link, —CH_2—, whereas most of the engineering polymers have a high concentration of large, voluminous ring structures, such as the aromatic phenylene ring, —C_6H_4— or the hexamethylene ring, —C_6H_{10}—. It becomes quite plausible to envision that, upon heating, these polymer chains become less likely to slide past each other. They will with steadily increasing temperature remain firm for a longer time but, on the other hand, must be processed at higher temperatures too. This effect can be illustrated very well with polycarbonate as an example. Regular polycarbonate contains two rings in their basic unit, whereas the high-temperature polycarbonate bears three rings.

Other resins obtain their engineering plastics status by employing hydrogen bonds. These are weaker than regular covalent chemical bonds but can form between adjacent polymer chains if they contain oxygen or nitrogen atoms as part of their chain links. Notable examples are nylon and acetal resins. The only way polyethylene can attain similar outstanding properties is by either increasing the molecular weight to the limit, such as the ultrahigh molecular weight polyethylene or by meticulously arranging the chains in an orderly crystalline structure, such as the superdrawn polyethylene fiber Spectra® (Trademark for AlliedSignal Inc.).

One group of engineering plastics is characterized by the presence of ester groups.

$$-\underset{\underset{O}{\|}}{C}-O-$$

They are formed by the chemical condensation reaction of hydroxyl and carboxyl groups containing substances. Due to their structure they readily react with water at molding temperature, forming low molecular weight entities. Therefore, they all have to be well dried prior to extrusion. Polycarbonate and all the thermoplastic polyesters belong into this class.

Polycarbonate (PC) distinguishes itself from other thermoformable plastics by its high impact strength over a very wide temperature range, its good clarity, and its higher glass transition temperature exceeding those of the polyesters. The limiting factors for its wider field of applications are its higher price and its susceptibility to environmental stress-cracking when in contact with certain solvents.

Polyethylene terephthalate polymers (PET) have been described earlier under orientation and crystallization. Their application for thermoforming has been greatly extended in recent years by increasing the molecular weight and by copolymerization. Products are now on the market, which display clarity equal to polymethyl methacrylates. The realm of polyesters has been extended considerably by the partial substitution of either of its two components, ethylene glycol or terephthalic acid. For packaging and industrial applications where strength at higher temperatures is required either the glycol component can be partially replaced by the ring-structured glycol, cyclohexanedimethanol, or the diacid component partially replaced by the two aromatic rings containing diacid 2,6-naphthalenedicarboxylic acid.

A great variety of engineering thermoplastics belong to the group of *polyamides* or generally called nylons. These polymers consist of neighboring amido —NH— and carbonyl —CO— groups, which are interspersed with short chains of methylene groups. The most common resins of this group are semicrystalline and have high melting points. Only a few grades are thermoformable since the molecular weight and melt viscosity of most grades are too low. Some toughened, alloyed, or amorphous extrusion grades have found more applications for thermoforming. They have toughness, good wear, and chemical resistance with temperature capabilities close to the polyesters.

Polyphenylene ether (also called polyphenylene oxide) polymers (PPE or PPO) have an excessively high processing temperature and are, therefore, only available in alloy form. The nylon-modified alloys have a higher temperature capability though somewhat lower stiffness than the (impact) polystyrene-modified form. Sheets of these materials are available in pigmented form and are well suited for thermoforming.

The corresponding sulfur compound is *polyphenylene sulfide* (PPS). It is the highest temperature-resistant semicrystalline thermoplastic resin, is flame retardant, and has excellent chemical resistance. The polysulfone and polyethersulfone are amorphous, clear plastics having still better thermal capabilities than polycarbonate. A still higher temperature plastic, the polyetherimide has gained application where flame and high-temperature resistance is required. Plastics in this last group are, however, restricted to few industrial and aviation applications due to their high price. Special related forming processes have been recommended for these or other engineering plastics or for specific applications.

Copolymers, blends, and alloys

Only rather simple monomers are commercially used to produce polymers, which, when blended with additives, are sold as plastics for a variety of processing methods. Although certain changes in their properties are achievable by varying the polymerization process conditions, e.g., low- and high-density polyethylene, etc., there is still a demand for materials possessing properties not realizable with any of the basic polymers (homopolymers).

By reacting two or more different monomers either concurrently or sequentially, an unlimited number of copolymers can be produced. Their properties can be widely diversified through selection of the monomers, their relative concentrations, and the timing of their addition to the polymerization process. Often, copolymers are more accurately described according to the distribution of the various monomers within them. These names outline satisfactorily their particular arrangement as alternating copolymers, random copolymers, block copolymers, and graft copolymers. The designation terpolymer is used for a number of copolymers containing three different monomers.

However, these reactions are limited to monomers that combine under similar conditions. It cannot be used, e.g., for a mixture of monomers where one reacts by addition and the other by condensation or oxidation. When it becomes desirable to combine such components and when it becomes cumbersome to conduct the chemical reaction repeatedly in a consistent way, first, simple polymers or copolymers are produced and later, in a second step, these are blended together at the desired ratio. This is usually performed when all other additives are compounded into the plastic. One material that should be mentioned here is the well-known acrylonitrile-butadiene-styrene terpolymer, which can be made by graft copolymerization as well as by blending two copolymers.

Blends of different polymers have also been called alloys, to signify the synergistic effect observed in certain properties of these compounds. In most cases, however, the properties of alloys lie somewhere in between the properties of their component polymers. This may also be true for their prices.

The properties desired in an alloy could be any one of the following: clarity, ultraviolet resistance, heat resistance, low-temperature impact resistance, strength, processability, and last but not least, decreased cost. Many copolymers, blends, and alloys have already been mentioned in this chapter, and a large number are listed in Appendix C under their proprietary names. Alloys have gained importance, given the broadening use of plastics for highly specific applications in large quantities.

Fiber-reinforced thermoplastics

Sheet-molding compounds have for decades found many applications where strong, prodigious parts had to be produced at low cost. The binder for holding the reinforcing fibers (in most cases, glass) together consists of unsaturated polyester resin that must be cured in a heated mold. In recent years a number of very long fiber-reinforced thermoplastic composites in sheet or roll form are being sold for similar applications. The main difference is that in the latter case the sheet must first be heated to at least 50°F beyond the melting temperature or the softening range of the resin before it is formed and solidified by cooling in a mold. This inexpensive forming process is often referred to as *thermoplastic composite sheet stamping*.

Polypropylene, being the first polymer used in such applications, is now being augmented with a large number of engineering polymers (both amorphous and semicrystalline) up to polyetherimide and polyphenylene sulfide, the material with the highest melt temperature. The mechanical properties obtainable are mainly dictated by the kind of reinforcing fibers, their orientation, and their concentration. Compounds that contain, besides glass fibers, also carbon or aramid fibers are now offered. A unidirectional layout results in highest strength parts, but, only in one direction. For most applications continuous random glass fiber mats or randomly distributed long chopped fibers are being used, mainly for processing reasons. If compared with unfilled, or just pigmented thermoplastic sheets, the higher the fiber content the higher forces must be provided for forming. Therefore, depending on the formulation used, matched-mold compression molding or ridge and drape forming processes prevail over vacuum and air pressure forming. Such materials are available from Azdel, Inc., DuPont Automotive Products and Quadrax Corp. and are listed in Appendix C.

Transparent materials

The optical clarity of materials is critical where thermoformed parts are intended either for glazing-type or for packaging applications. Naturally, any transparent material can be made opaque by the addition of pigments. A small amount of pigmentation renders a material translucent, which means that although much of the light will pass through it, objects behind it are concealed or cannot be seen clearly. Thin films of translucent materials may appear transparent as long as the object is placed in contact with it. Surface roughness will also affect the clarity and brilliance of transparent materials.

The table below lists some of the naturally transparent and translucent materials:

Transparent Plastics	Translucent Plastics
Acrylic	Polyethylene
Cellulose acetate	Polypropylene
Cellulose propionate	Polyallomer
Cellulose acetate butyrate	High-impact polystyrene
Ethylene-vinyl acetate copolymer	Acrylonitrile-butadiene-styrene terpolymer
Polymethylpentene	
Ionomer	Polyvinylidene fluoride
Polystyrene	Polyamide
Styrene-butadiene copolymer	Fluorinated ethylene propylene copolymer
Styrene-acrylonitrile copolymer	
Polyvinyl chloride	Polyethylene terephthalate
Vinylidene chloride copolymer	Polybutylene terephthalate
Polycarbonate	Polyarylketone
Polyethylene terephthalate	
Polysulfone	
Polyarylsulfone	

A number of commonly opaque polymers have been made available also as transparent or nearly transparent materials. Because opacity is based on the presence of crystalline phases, transparency can be obtained by preventing the formation of crystals, e.g., the addition of another comonomer or the rapid cooling and stabilizing of the random polymer chain segments by cross-linking or orientation. If the size of the crystallites can be restricted to a size below the visible wavelength, the plastic will also appear clear. This can be achieved through the addition of microcrystalline nucleating agents.

Barrier materials

For the food-packaging industry the barrier properties of the plastic material are of great importance. Because glass and metals are totally impervious to any other substance, they have provided ideal packaging materials for centuries. However, their high cost and weight have limited their use. When lower cost plastics became available, they were tried as replacements. But it became evident that all organic substances and, therefore, all thermoformable plastic film and sheet materials permit gases (oxygen, carbon dioxide, water vapor, etc.) and liquids (water, flavoring agents, etc.) to permeate. A two-way process, permeation can lead to eventual deterioration of the quality and wholesomeness of packaged goods. Mainly, the volatile flavoring substances or important food taste ingredients may get lost into the environment or just getting absorbed by the packaging material, generally described as aroma scalping. Concurrently, the plastic material must not impart any taste or odor on its part to the content.

Two material properties dominate the permeability of gaseous or liquid substances through plastics materials. One is the solubility coefficient (S) of these substances in the plastic film materials and the other the diffusion coefficient (D) for the same pair of materials. The permeability (P), which according to the equation

$$P = D \times S$$

is defined as the product of diffusivity and solubility, gives the useful value characterizing the transport rate at a steady state.

The permeability of gases is dominated by the diffusivity because their solubility in the plastic is negligible. Carbon dioxide occupies a unique position since it can become quite soluble in plastics at high pressures. In general gas molecules just pervade the "free volume" of the polymer and migrate from spot to spot, depending on the partial pressure on each side of the film. The speed is mostly determined by the size (molecular weight) of the gas.

The solubility of liquids in plastics materials is very much dependent on the chemical composition. The hydrophilic resins (polyvinyl alcohol, nylon) are capable of absorbing appreciable amounts of water or water vapor, and some of the hydrophobic plastics (polystyrene but mainly the polyolefins) show an affinity to organic solvents (gasoline) and oils (aroma substances). An additional effect must always be kept under consideration, that the inclusion of only a few percent of such liquids can greatly affect the plastic's mechanical properties too. All solvents, including water, act like plasticizers if absorbed by the plastic.

Besides the chemical structure of the polymer chain, the morphology, the degree of polarity dictating interchain forces, the chain stiffness, the chain orientation, and the crystallinity fraction all have an effect on permeability. The crystalline phase is regarded as impermeable, but due to the stresses present at the crystal-amorphous interphase, high crystallinity does not necessarily result in low permeability. This may also occur when crystallinity is being enhanced by annealing. Liquid crystal polymers have very low permeabilities, but they are not available in film form and are, therefore, restricted to polyester alloys containing only 10% LCP.

Orientation can also give either rise to permeability or diminish it. The drawing of semicrystalline plastics can decrease permeability by up to two orders of magnitude. In general quenched barrier films are subjected to stretching to avoid crystal formation. On the other hand, excessive, especially biaxial orientation, may lead to the formation of microvoids. It should not come as a surprise that in the case of poly(4–methyl-1-pentene), where the amorphous phase has a higher specific gravity than the crystalline one, these rules might just be reversed. Because not enough is known about all the quantitative influences, it is still necessary to check out the suitability of film materials for each application.

The permeation of moisture through the package into the food can make crispy snacks unappealing. The permeation of oxygen into the packaged materials can turn fatty foods rancid, but in other cases the permeation of oxygen becomes desirable when the fresh red color of meat should be prevented from turning brown in the packaging.

If the correct barrier properties have been selected, the amount of a substance migrating through the package in any direction within its expected storage time span should be negligible. Because many food products are stored only for a limited time and because different foods have different sensitivities in regard to permeable substances, the plastic composition of the film must be matched to the specific application. In Table 9.1 the key barrier properties of various basic plastic materials are compared. These values are strongly dependent on temperature, thickness, possibly occurring voids, concentrations, relative humidity, and other environmental and testing conditions. For convenience the numbers for 1-mil-thick films are expressed for gases in cm^3 at a film thickness of 1 mil/100 sq in., 24 hours, 1 atm at 73°F and for water vapor in g at a film thickness of 1 mil/100 sq in., 24 hours at 90% R.H. and 100°F. The numbers are meant for purposes of comparison. Too many variabilities must be taken into consideration when factual numbers must be established. In all cases leading to practical applications the selected combinations must be proven in tests under applicable conditions. It must not be overlooked that good mechanical properties and processability

TABLE 9.1.
Barrier materials properties.*

	Gas transmission rate** at 1 atm and at R.T. $\frac{cm^3 \cdot mil}{100 \; in.^2 \cdot 24 \; hr \cdot atm}$			Moisture vapor transmission rate** at 100°F, $\frac{g \cdot mil}{100 \; in.^2 \cdot 24 \; hr}$
	Oxygen	Nitrogen	Carbon dioxide	
Polyacrylonitrile (PAN), cast film	0.01	—	1	1
Ethylene vinyl alcohol copolymer (EVAL)	0.05	—	0.1	4
High-barrier polyvinylidene chloride copolymer (PVDC)	0.1	0.1	1	0.1
Acrylonitrile methyl acrylate copolymer	0.6	0.2	2	4
Regenerated cellulose, surface coated (Cellophane)	0.6	1	0.5–5	1–10
Polycaprolactam, polyamide (Nylon 6)	2	1	10	20
Polyethylene terephthalate, oriented (OPET)	5	1	20	2
Polychlorotrifluorethylene (PCTFE)	10	2	20	0.03
Polyvinyl chloride (PVC)	15	20	30	5
Cellulose acetate (CA)	100	30	1000	20
Polypropylene (PP) and high-density polyethylene (HDPE)	150	30	500	0.5
Polycarbonate (PC)	300	50	1000	10
Polystyrene, oriented (OPS)	300	50	1200	8
Fluorinated ethylene propylene copolymer (FEP)	750	300	1500	0.4

*Refer to Appendix D for the conversion factors for listed values.
**All film materials listed—with the exception of the first one, polyacrylonitrile cast film—are commercially applied packaging films. They are ranked in order of increasing oxygen permeability. This ranking is nearly paralleled by their permeability to other gases. A high water vapor transmission rate can affect the gas permeability of films under high-humidity conditions. Since published data vary widely, the values in the table are averages.

must also be met by the film material to become useful as a packaging material. One of the oldest and still low-cost film material—the surface-coated and biomass-based cellophane—would still present a preferable barrier material if it were stronger (especially in tear resistance) and could be formed.

As a matter of fact, the barrier properties of some materials are so efficient that only a very thin layer of them is needed to satisfy barrier requirements. A thin coating of silica, which is applied by a low-temperature plasma chemical vapor deposition in vacuum, can improve barrier properties of transparent films several hundredfold. The Simplicity® process is provided by PC Materials, Inc. (Pennsville, NJ) and the QLF™ (Quartz-Like Film) barrier coating by BOC Coating Technology (Concord, CA). Similar properties are achieved by thin aluminum coatings, applied by vacuum metallizing, yielding a highly reflective film. In some cases a coating is applied after packaging materials are thermoformed. In others a barrier material is formed just on the surface by means of highly reactive chemicals. This can be accomplished according to the "Airopak" process (Air Products and Chemicals, Inc. Emmaus, PA) by contacting the formed part with a nitrogen-diluted fluorene gas at ambient temperatures. At one time sulfur trioxide (2% concentration) was also used to introduce sulfone groups to the plastic's surface. After either treatment high-density polyethylene parts will not pick up solvent molecules and thus no paneling will occur, which means swelling and buckling of the container.

By thorough investigations the ideal packaging material for most big-volume food products have already been established. In most cases these are laminates and coextrusions, which combine one or two barrier materials, a tough mechanical film component and a material that will provide good heat sealability to at least one surface layer. In some cases additional adhesive layers must be provided if the bond between the functional layers is not adequate. Because the capacity to impede gas permeation of the best barrier polymers deteriorates appreciably in the presence of moisture, some laminates contain, besides a moisture barrier layer, also an additional moisture-scavenging layer to absorb the small amount of moisture that might penetrate the protective layer. For oxygen-sensitive goods the oxygen absorber Ageless from Mitsubishi Gas Chemical Co. can be incorporated into the packaging film. The same principle is utilized by Japan's Toyo Seikan Group in their Oxyguard material. The specially reduced iron compound in one of the layers is capable of scavenging the oxygen that could not be removed by gas flushing or vacuum packaging or that later diffused through the barrier. Recently, a product of undisclosed composition has been offered on the market by Amoco Chemicals (Chicago, IL) under the name of Amosorb.

If the water-containing content requires a steam sterilization process, it might become necessary to select a somewhat higher water permeable outer layer, e.g., by replacing the polypropylene film by a polycarbonate film layer. Its water permeability at sterilization temperature is about equal, but at ambient temperatures it is 10 times larger than that of polypropylene. Such a structure facilitates after retorting the drying out of the intermediate polyvinyl alcohol layer.

The best barrier materials cannot be extruded into a film material unless they are copolymerized with other monomers. The neat polyacrylonitrile, polyvinylidene chloride, and polyvinyl alcohol films can be obtained only by a costly solution-casting process, whereas the corresponding copolymers can be inexpensively converted into a film material by an extrusion process that becomes even more attractive when several layers are produced simultaneously in a coextrusion process.

Some common three-layer coextruded films are:

(1) High-density polyethylene inside low-density polyethylene to increase stiffness and reduce loss of moisture
(2) Acrylonitrile copolymers inside polypropylene to produce gas barrier properties
(3) Ethylene-vinyl alcohol copolymers inside nylon or polypropylene to obtain gas and aroma barrier properties
(4) Vinylidene chloride copolymers inside high-impact polystyrene or polyolefins to obtain gas and moisture barrier properties
(5) Nylon within low-density polyethylene to obtain gas and moisture barrier properties.

As some of the distinct packaging structures just two examples should be cited:

(1) For blister packaging of pharmaceuticals 2-mil polychlorotrifluoroethylene, 2-mil low-density polyethylene, and a 7.5-mil polyvinyl chloride laminate, each bonded with a thin adhesive layer, are recommended.
(2) An old competitive material, the regenerated cellulose film cellophane, coated on both sides with nitrocellulose lacquer (the oldest synthetic multilayered packaging film) should be mentioned as the first moisture-resistant, gas-, and aroma-barrier film.

The structuring of barrier films has been enlarged by increasing the number of layers to five and seven. The development of special materials for packaging applications is still progressing, and certain shifts in market domination—partly initiated due to customer preferences—have already taken place during the last 10 years. The following product names (old and new) are listed alphabetically along with their manufacturer and chemical composition (if commonly known):

Name	Company	Description
Adheflon	Elf Atochem	Polyvinylidene-fluoride-based (PVDF) tie-layer resin for nylon 12 to PVDF
Barex	BP Chemicals	Acrylonitrile-methyl acrylate copolymer, rubber modified
Bynel	DuPont Packag.	Coextrudable ethylene copolymer tie-layer for multilayer barrier structures
Cotie	BP Chemicals	Coextrudable tie-layer material for Barex
EVAL	EVAL Comp.	Ethylene-vinyl alcohol (60 to 75 mol%) copolymer
Exceed	Okura Industrial Co.	Biaxially oriented, high barrier polyvinyl alcohol film
Lamal	Morton International	Urethane prepolymer adhesive
Lamicon	Toyo Seikan	Ethylene-vinyl alcohol copolymer, laminated within nylon or polyolefin films
Modic	Mitsubishi Gas Chemical	Coextrudable grafted polyolefin tie-layer material
MXD6	Mitsubishi Gas Chemical	Aromatic polyamide high-barrier resin
Mylar M	DuPont Films	Polyvinylidene chloride coated (both sides) on polyethylene terephthalate film
Plexar	Millennium Petrochem.	Coextrudable ethylene copolymer tie-layer for multilayer barrier structures
Saran	Dow Plastics	Vinylidene chloride-vinyl chloride copolymer film
Saran F	Dow Plastics	Solvent-soluble vinylidene chloride copolymer for high-barrier coatings
Selar PA	DuPont Films	Amorphous nylon resin (hexamethylene diamine and iso/terephthalic acid polymer) additive for ethylene-vinyl alcohol copolymer for solid phase pressure forming
Selar PT	DuPont Films	Modified polyethylene terephthalate copolyester to produce by "Fortex" thermoforming process a clear monolayer, high-temperature barrier container

Selar RB	DuPont Films	Laminar blend of nylon or ethylene-vinyl alcohol with polyolefin for packaging
SIS	Shell Chemical	Coextrudable styrene-isoprene-styrene block copolymer tie-layer material
Soarnol D	Morton International	Ethylene- (29 mol%) vinyl alcohol copolymer
Unite	Aristech Chemical	Polyolefin with anhydride functionality

It is evident that for applications where mass production and automation are so important, the development of suitable machinery and an efficient process is of equal importance. For some of these the reader is referred to Chapter Twelve, "Related and Competing Forming Processes."

The usage of plastics in packaging has made great inroads during the past decade and is expected to expand further in coming years. Besides performance attributes the packaging industry is much affected by appearance, physical size and shape, and customer preferences. Therefore, many changes for the future should be anticipated, including efforts on developing workable recycling processes. Some of the possibilities are to improve the compatibility of the various multilayer components or the mechanical separation of incompatible components.

Electrical properties

Thermoformed parts are frequently used for applications where their electrical properties are of prevailing importance. All plastics are generally excellent insulators. The minor differences in surface and volume resistivity among them can be ignored in all but a few special applications.

Plastics surfaces will easily become electrically charged when in contact with other insulators or may already be highly charged when removed from the mold. Dust particles are attracted to exposed surfaces, and delicate electronic components could be damaged when encountering such charges in plastics packaging. To obtain an electrostatic discharge (ESD) characteristic, only a small amount of an additive or surface treatment is required. Most of them work on the principle that they attract a small amount of moisture to the surface. This explains that its effectiveness depends very much on the relative humidity in the environment and that the protection may wear off in time.

Though electrically conductive plastics are not yet available, additives or surface treatments are recommended where a small degree of conductivity is required. The extensive range in electrical conductivity can be divided the following way:

Surface Resistivity of Various Materials

Material	Surface Resistivity (Ω/square)
Neat plastics	$10^{12}-10^{18}$
Antistatic plastics	$10^{8}-10^{12}$
Static dissipative plastics	$10^{3}-10^{8}$
EMI shielding plastics	$10^{0}-10^{3}$
Semiconductors	$10^{-5}-10^{7}$
Metals	$10^{-5}-10^{-6}$

Where thermoformed parts are used as covers for high-frequency electrical and electronic equipment, and where either such equipment inside the enclosure must be protected or radiation to the outside prevented a much higher electrical conductivity, than that sufficient for electrostatic discharge, must be obtained. Therefore, for the prevention of electromagnetic interferences (EMI) and radio frequency interferences (RFI), the plastic materials must be compounded with conductive fibers, either metallic fibers or carbon and graphite fibers mostly electrocoated with nickel. Otherwise, a continuous metal coverage of either surface of the part can lead to a satisfactory protection. This can be achieved by a coating with a conductive paint, by vacuum metallizing or by electrolytic or electroless metal plating, all involving expensive post-mold procedures.

The 3M Company (Austin, TX) has, therefore, developed a sheet material (0.020" thickness) that can be laid on top of many structural plastic sheets and together thermoformed by vacuum or air pressure to the desired shape. It consists of a network of metal alloy fibers on top of an ethylene-vinyl acetate (EVA) polymer layer. Both become stretchable at the thermoforming temperature and will bond like a hot-melt adhesive to the substrate. Because the metal fibers melt at 281°F during thermoforming and become liquid, a forming process must be selected where this layer will neither come in contact with the mold nor the plug surface.

Plastics recycling

There is no question that the recycling of plastics is becoming increasingly important. The burying of used plastic parts is definitely undesirable because that represents a waste of a natural supply resource. Most plastics used to be made out of coal but now are based on crude oil and natural gas, which have been taken out of the ground, purified, and converted under expenditure of energy and application of various chemical processes into a useful product.

No problems should exist with the reuse of factory generated waste, such as sheet and film edge trims generated in the extrusion process as long as the regrind is kept clean and well mixed in a constant proportion into the virgin material. If the thermoformer has to collect the trim margins and ship it back to the company extruding the sheet, one must already accept an inferior quality product due to possible contamination. The reprocessing of post-consumer generated waste will—with just a few outstanding exceptions—under the best conditions yield a material suitable for extruding plastic wood.

Great efforts are being made to reutilize the material extensively used for beverage containers. The stretch-blow molded polyethylene terephthalate, carbonated soft drink bottles are the most suitable plastic for recycling. They are available at a very high volume, though consisting more of air than plastic. They can easily be separated from other waste streams and exist only in two colors, clear colorless and clear green. Still, they must be, at considerable cost, collected, separated from labels, bases, and metal closures, and then shredded, cleaned, and dried. Polyvinyl chloride, which could be a look-alike, must definitively be excluded. Due to possible contaminations the polyester as such cannot be reused for food-contact packaging applications. The most promising use for them is the conversion to carpet textile fibers if the molecular structure of the polymer should be preserved. The next prospect for their reapplication consists of regaining some of the chemical components by chemically reforming the large polymer molecules into smaller molecular entities via a thermal cracking, distillation, hydrolyzation or transesterification process. Still, none of the reclaiming schemes have resulted in an economically viable process. They all are maintained only by government-imposed regulations or consumer taxation, such as the cost for separate collection, etc.

Other plastics, whether consisting of discarded milk or detergent bottles, other types of packaging or structural parts (from junked automobiles, appliances, instruments, or equipment), do not fare as well. The multitude of compounding ingredients and mix of polymers complicate the reprocessing of the great number of plastics materials. The polymers of many of these parts might already have been degraded due to ultraviolet radiation, heat, or contact with chemical agents. This would be especially true for parts that already contain appreciable amounts of regrind. Thermoplastic resins will, after several reprocessing steps, turn into quite brittle or dark-colored masses.

This stands in contrast to the reuse of glass or metal containers since these materials retain, after remelting, their properties without changes.

Due to the low weight and convenience of plastics packaging, they have already contributed to energy savings over those other materials. Because all plastics at one stage originated from fuel materials (natural gas, crude oil, or coal), one should not object to redirecting them back to that type of material at the end of their life cycle, to regain at least their intrinsic value as fuel. For these waste-to-energy processes only a limited effort for segregation and cleaning is required because the other commingled organic materials, such as paper, cardboard, rubber, and food residues, also contribute to energy gains. With the exception of polyvinyl chloride and a few highly compounded plastics, all plastics will generate during combustion in air twice as much heat as the biomass materials, such as paper, cardboard, and wood.

The fears that toxic ingredients could be liberated into the environment should be alleviated by significant reductions in the use of toxic stabilizers and pigments, a trend which, at this time, is being pursued for many other reasons. Besides, ashes and gaseous combustion products are already made subject to the separation and chemical bonding processes and are increasingly utilized by coal-burning power plants.

To find alternatives for land-filling large amounts of waste plastics, the chemical industry has pursued the development of biodegradable plastics. Thirty years ago one could hear people saying, "Soon the world will be covered in Saran," based on the belief—at the time—that polyvinylidene fluoride can withstand unharmed 50 or more years of outdoor exposure. Now it is known that most thin films of thermoplastics, so widely used in packaging, will disappear within one year or so. In order to make them break down within one month, like paper and cardboard does, film products containing at least 15% of organic materials, in most cases cornstarch, are on the market. It must be mentioned here that plastics will survive in landfills for decades, but so might also newspapers, textiles, and food leftovers.

Several companies promote the utilization of inherently biodegradable polymers that contain in their chain such monomer units that can be classified as nutrients for microorganisms or that are especially sensitive to UV radiation. Into the first group belong the polyhydroxybutyratevalerate polymers and the polylactic acid based polymers. The latter is marketed by Cargill Inc., Minneapolis under the name of EcoPLA. Products of the latter group are obtained by incorporating carbonyl groups during the polymerization of ethylene. However, the considerably higher cost of these materials hampers their broad acceptability.

Flammability of plastics

The chapter on materials cannot be closed without some remarks in regard to the behavior of plastics in case of fire. Plastics consisting of organic chemical materials are subject to violent oxidation reactions in the presence of air at high ambient temperatures, like any other organic materials, e.g., wood, paper, textiles, and dry food products.

Extremely flammable plastics, such as nitrocellulose, have long been outlawed. Modern plastics are about on a par with the above-mentioned organic, natural materials. However, due to many other circumstances, the likelihood of plastics becoming involved in fires seems to be higher. Therefore, many plastics are now available compounded with flame retardant and smoke suppressant ingredients. Because these contribute to higher costs, impair some properties, and make processing more difficult, they are used only when necessary.

The number of laboratory tests for the determination of flammability has multiplied substantially during the last decades. Still, they cannot duplicate all the possible fire scenarios. This becomes clear to all those who want to start a fire without resorting to gasoline. For the beginning, the fuel must be available in a sufficiently thin form (paper, film, or small sticks) to raise its temperature beyond its thermal decomposition point. Flame spread depends very much on geometrical arrangement of the flammable materials. Last but not least the fuel value, the amount of heat that can potentially be released, must be considered. As a general rule for plastics it can be stated:

(1) Polymers with a high content of hydrogen atoms exhibit higher flammability than those with a low content. These two extremes are represented by polyethylene on one side and the highly aromatic ester, ether, ketone, or sulfone polymers on the other.

(2) Because both hydrogen and carbon atoms contribute to fuel value, the polymers that contain oxygen, nitrogen, and especially chlorine in larger amounts in their structure will generate less heat during combustion. As example the acetal resins, polyvinyl chloride, and also highly mineral filled plastics should be listed.

More worrisome are the thermal decomposition products of any material—not only those containing plastics—under incomplete combustion conditions, because they contribute to the formation of smoke and toxic substances. Under those circumstances potential lethal episodes have always been traced to carbon monoxide, no matter what the composition of the fuel source was.

There is very little any one of the plastics processors can do to remedy this situation. Due to their generally thin film or sheet thickness, thermoformed products are more vulnerable to inflammability and flame spread than other more massive plastic parts. On the plus side, the light weight of thermoformed parts contributes less fuel.

In many cases legislative rulings will dictate which of the various government-specified tests must be passed. Submitting samples for testing to an independent test laboratory is essential. In cases where specific rules are not provided, one should still be cognizant of potential hazards surrounding the application of the formed part and at least communicate with the materials supplier to ensure that the most suitable formulation for a particular application is secured.

Toxicity of plastics

The high molecular weight polymers constitute the major component of common plastics materials. They are basically nonreactive under ordinary environmental conditions. Most of them can be used for food-packaging applications, and some of them even find uses as body implants.

That nearly all plastics are made of highly chemically reactive, low molecular weight compounds commonly called *monomers,* which as such are highly toxic, does not pose a problem to the plastics user. These materials remain contained in sealed autoclaves and pumping systems. Decades ago when minute amounts of unreacted monomers were detected in plastics, they were eliminated from all plastics intended for food-contact applications. Other materials, including catalysts, emulsifiers, solvents, etc., are added to perform the polymerization, and additional materials, such as stabilizers, lubricants, pigments, fillers, etc., are introduced to the polymer during compounding. Some of those that could be considered toxic have been or are in the process of being eliminated; the overwhelming number of them are considered innocuous. Only food grade plastics, which are now widely available, should be employed when food contact or extended skin contact can be anticipated.

As indicated, a chemical breakdown of high molecular weight materials will occur in case of fire. A similar decomposition may also occur at elevated temperatures, especially if recommended processing temperatures are inadvertently surpassed. No matter whether the polymer reverts to the monomer or whether it forms other low molecular weight gases or volatile reactants, a toxicity problem may arise. Vinylchloride-containing polymers may, under those conditions, form a corrosive acid. Problems have also been caused in the packaging field by improper heat-sealing conditions.

TEN
Thermoforming Processes

MANY THERMOFORMING PROCESSES have evolved from the first forming method where live steam was used both for heating and forcing flat thermoplastic sheets (camphor plasticized nitrocellulose) into a contoured mold. The various ways of preparing a plastic for the actual forming and most of the machinery for accomplishing the forming have been discussed in detail. Here, only the proper selection and sequencing of the forming of a sheet or film material will be described.

Starting from the basic forming processes employing pneumatic pressure and three simple forming tools—a clamping frame, a concave or female mold, and a convex or male mold—we shall examine increasingly sophisticated installations. All such variations were developed to offset the shortcomings of these less proficient, basic methods.

In the series of drawings to follow, the illustrations commence with an already heated plastic sheet (not showing the heaters) and concentrate on the movements taking place during forming.

Billow, bubble, or free forming

Some thermoformed shapes can be obtained without the use of a mold. In billow forming the uniformly heated sheet is sealed to a plenum chamber (also called vacuum and pressure box or table). If either vacuum or air pressure is admitted to the chamber, the pressure difference will cause the sheet to bulge inward or outward. To repeatedly obtain the same shape, the amount of air or the pressure must be controlled for each piece to compensate for slight variations in temperature. Figure 10.1 depicts an arrangement in which a light beam controls a valve that regulates the depth of draw. Aside from the depth, only the shape of the perimeter can be varied; therefore, part shapes are limited. Figure 10.2 depicts one part formed with only a flat opening and another formed with a three-dimensional opening in the box. These formed parts have useful optical

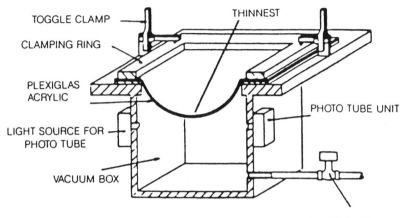

Figure 10.1. Free forming with vacuum (courtesy of Rohm and Haas Co., Philadelphia, PA 19105).

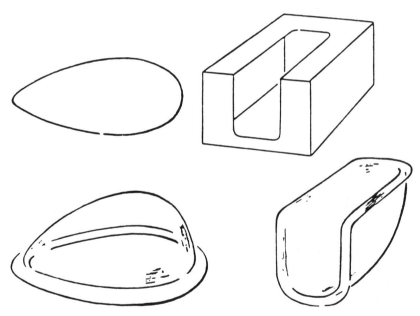

Figure 10.2. Free formed shapes obtained with a flat and a three-dimensional opening (courtesy of Rohm and Haas Co., Philadelphia, PA 19105).

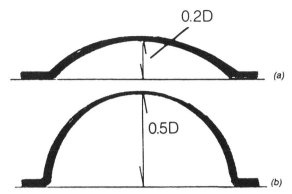

Figure 10.3. Material thinning at apex of spherically blown shapes: (a) height to diameter ratio 0.2, apex thickness 80%. (b) Height-to-diameter ratio 0.5, apex thickness 33%.

properties, including transparency with minimal optical distortion, because the soft sheet does not touch any mold surface. Unfortunately, cooling of the sheet proceeds at a very low rate.

It is commonly assumed that the wall thickness of a blown shape will be of uniform thickness as is observed when inflating a rubber balloon. However, this is only true for cross-linked polymers. Thermoplastic materials thin out nonuniformly: the area in the center becomes the thinnest and the area near the clamping frame the thickest. Because small temperature differences affect the modulus of elasticity, unevenly heated sheets cannot be formed by this method. In fact, most thermoplastics produce oddly shaped forms that are easily punctured. Only three transparent materials, due to their tenacity and thermoelastic behavior at elevated temperatures, are appropriate for this technique: polymethyl methacrylate, the cellulosics, and polycarbonate. Figure 10.3 points out the best material distribution obtainable with free-formed spherical shapes. Because of excessive reduction in thickness at the apex, a height-to-diameter ratio of 75% appears to be the practical limit.

Cavity forming

This process, also called straight or simple vacuum forming, utilizes a female, negative, concave, or cavity mold. The clamped and heated sheet is brought into contact with the edges of the mold to obtain a seal. Atmospheric air pressure will press the sheet against the cavity surface when the space between cavity and sheet is evacuated.

This simple process, as shown in Figure 10.4, is suitable for relatively shallow parts, since the sheet will thin out more and more, the deeper the

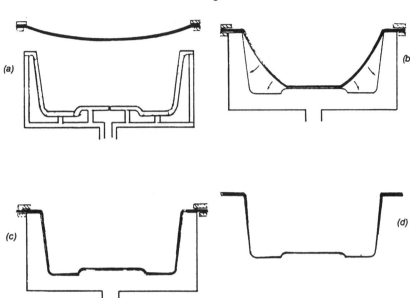

Figure 10.4. Cavity forming: (a) female mold with heated sheet, (b) and (c) vacuum applied to mold, (d) cavity formed part.

cavity. Otherwise, the process is ideal for fabricating parts with a sturdy frame, because the wall thickness at the frame or edges will remain close to the original sheet thickness. The thinning of contoured parts is particularly noticeable on sharp inside corners, which, therefore, should be avoided.

Drape forming

The third basic thermoforming method, shown in Figure 10.5, utilizes a male, positive, convex, or drape mold. The sheet is again clamped in a frame. When the material is sufficiently heated, the frame is lowered until the sheet seals to the vacuum box. Because the mold projects toward the sheet's plane, the sheet will first contact the elevated areas of the mold and start to solidify there at nearly the original thickness of the sheet. The remaining area of the sheet is stretched during its downward movement, and further thinning occurs when the vacuum quickly causes the sheet to contact all mold surfaces. Here, too, the areas formed last become the weakest areas of the part.

In some cases the material's distribution can be marginally improved by inverting the mold and mounting it to the top press platen. The sheet will sag downward during heating and contact the crest of the mold at a later time, thus reducing the part's thickness in that area.

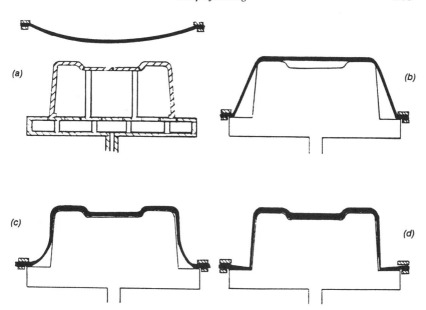

Figure 10.5. Drape forming: (a) male mold with heated sheet, (b) and (c) vacuum applied to mold after frame was lowered, (d) drape formed part still on mold.

If multicavities are to be used for cavity forming, no special provisions are required. In drape forming, however, the upper frame member should consist of a frame and grid to pull down the sheet to the vacuum box around each individual mold. Otherwise, the sheet might form a web between adjacent molds (see Figure 10.6). Since conveyor-fed thermoformers have no upper frame, a grid assist must be used, which is mounted to the top movable platen.

Many shapes can be produced either by the cavity or drape forming process, thus enabling one to pinpoint where to have the greatest material thickness. Inside corners of a female mold correspond to outside corners in a male mold. However, many other aspects, such as surface finish and mold shape, may also dictate which type of mold to use (See Chapter Three on molds).

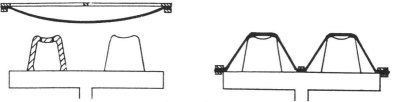

Figure 10.6. Multiple drape molds showing how bar assist helps drape sheet over each individual mold.

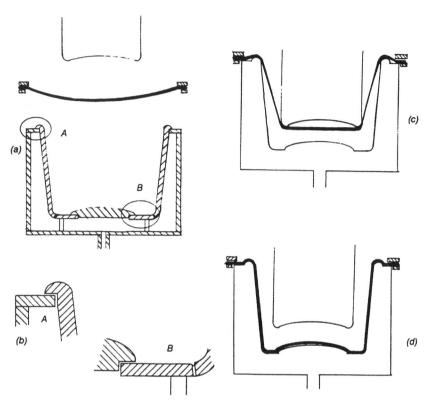

Figure 10.7. Plug-assist forming with cavity mold: (a) heated sheet inserted between mold and plug, (b) details of mold showing efficient air removal through slots and grooves, (c) mold raised and plug inserted, (d) vacuum applied to mold and part formed.

Plug-assist forming

Because in general a balanced distribution of the material over the whole part is desirable, prestretching the sheet prior to contacting the cooled mold surface is advantageous. There are several techniques for doing this. Naturally, this additional step prolongs production time and machinery investment expense.

In plug-assist forming, illustrated in the drawings of Figure 10.7, the shortcomings of the cavity and drape forming processes cancel each other out. When using a female mold, the bottom part would become the thinnest; therefore, a male plug should first contact the heated sheet without cooling it and drape it until the plug is close to the bottom of the cavity mold. Only then is the air between the female mold and the sheet evacuated to accomplish the final cavity forming step.

Billow drape forming 165

The plug should neither mark nor cool the heated sheet during contact. Therefore, it must have a very smooth surface, and if the plug is made out of metal, it must be heated slightly below the sheet temperature. Materials with low thermal conductivity, such as syntactic foam, coated wood, cast or machined thermoset plastics (particularly when covered with felt or velvet), will not need temperature control.

For this and the following processes sufficient machine opening between the platens is required. The opening must be at least 3 times the height of the formed part.

Billow drape forming

If one wants to use a male mold for producing the same part, a balanced distribution of the material can be obtained by utilizing another modification supplementary to that used in plug-assist forming. One of these processes is called billow drape forming or reverse draw forming (see Figure 10.8) and the other process is termed vacuum snap-back forming.

The heated sheet is first sealed to the plenum chamber and a bubble is

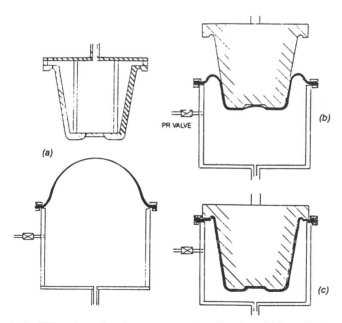

Figure 10.8. Billow drape forming or reverse draw forming: (a) heated sheet sealed to box and billow formed, (b) drape mold inserted while excess air bleeds through pressure relief valve, (c) mold fully inserted, vacuum may also be applied to mold.

extended outside the box by air pressure. This will stretch the sheet, resulting in a spherical shape thinnest at the apex. When the male mold is lowered, it will first contact the thinnest part of the sheet at the apex of the bubble. On further closing, some of the air must be vented, as the bubble decreases in size and eventually reverses itself completely. The final forming of the article is again done by a vacuum drawn through the male mold.

Snap-back forming

This forming method, sketched in Figure 10.9, is closely related to the preceding one. The only difference is that the bubble is first formed by vacuum and, therefore, is concave instead of convex. This process minimizes chill marks or mark-offs because no plug is used. The sheet does not have to be drawn the full height of the male mold if the draw ratio is high. After inserting the male mold, the vacuum in the plenum chamber is released, and the material is allowed to snap back. A vacuum must be applied to the male mold to complete the forming step, and/or the pressure in the box must be raised.

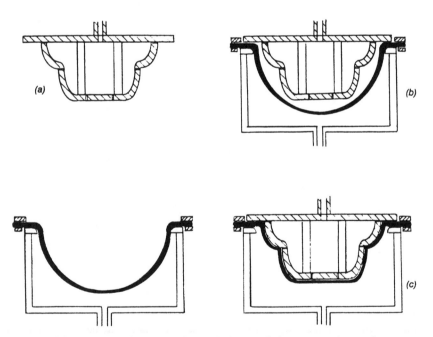

Figure 10.9. Snap-back forming: (a) heated sheet sealed to vacuum box and concave billow formed with vacuum, (b) male mold inserted, sealing sheet at the edges, (c) vacuum released in box which may become pressurized and/or vacuum applied to mold.

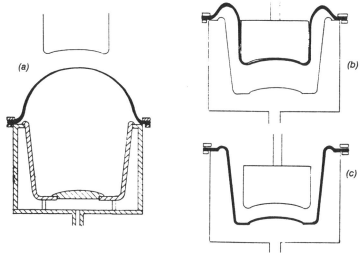

Figure 10.10. Reverse draw with plug-assist forming: (a) female mold box pressurized to form billow, (b) plug descending into billow while some of the air is bled, (c) plug at end of stroke and vacuum applied to mold.

Air slip forming

Air slip forming may be regarded as another variation of this forming technique. Instead of stretching the sheet in a billow shape, the male mold is inserted into the heated sheet while contact with the sheet is prevented by blowing preheated air through the bottom part of the mold, which makes the first contact with the sheet. Thus, the sheet does not become restrained there and continues to stretch uniformly until the male mold is completely inserted. At this point the flow of air is turned off, and a vacuum is applied as usual. This process requires exceptionally precise tuning of sheet temperature and vacuum and pressure settings.

Reverse draw with plug-assist forming

Combining several of the described techniques leads to reverse draw with plug-assist forming, one of the most sophisticated vacuum forming methods. In this process the wall thickness is completely controlled. The part's cross sections can be made fully uniform throughout or strengthened in any area. Adjustments are made by judiciously varying the size of the blister, the penetration of the plug, and the temperature of the plug.

In the reverse draw with plug-assist forming method (Figure 10.10), a bubble is first blown upward. Instead of a male mold, the bubble initially touches the male plug, or if the air slip method is included, sliding on the

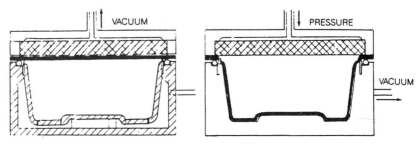

Figure 10.11. Trapped sheet forming. Left: sheet held by vacuum to porous heater plate; O-ring seals sheet to mold. Right: heated sheet formed by vacuum and/or pressure and knives will trim formed part when press is closed further.

air cushion in front of this plug. The billow will be reversed in a folding or rolling manner while the plug is lowered [Figure 10.10(b)]. After the plug reaches its final position, located at a set distance from the female mold, a vacuum is applied to the female mold, and the forming and cooling of the article are completed.

Trapped sheet pressure forming

Another thermoforming process that has captured a very wide application is the trapped sheet pressure forming process. It is the method of choice for thin, biaxially oriented materials, such as polystyrene or the polyolefins. If highly oriented sheets are clamped only in frames and heated, the liberated retractile forces may easily lead to excessive thinning or rupture of the sheet in certain areas. In the trapped sheet pressure forming process (Figure 10.11), the sheet is primarily held by vacuum to a porous heater plate heating the thin sheet rapidly and uniformly. The edges of the sheet are restrained to the female mold. The sufficiently heated sheet is rapidly depressed into the cold mold when the vacuum is switched to air pressure. By using an elastic seal and equipping the molds with steel knives, trimming can be accomplished through a brief additional exertion of pressure. This process represents the fastest thermoforming method.

Twin-sheet forming

For twin-sheet forming two sheets of plastic must be heated. They are then inserted together between an upper and lower female mold half, as illustrated in Figure 10.12. The pressure of the closing molds will weld at the circumference the two sheets together. An opening or hollow needle allows air to enter. Forming can be done either with compressed air or just by simultaneously evacuating both mold halves. The formed parts resemble parts manufactured by blow molding or rotational molding and can combine versatile textured and colored decorative effects.

The twin-sheet thermoforming process was developed for the fast production of large and rigid parts. These could be double-walled doors, extended containers, large machine covers, shipping pallets, customized dunnage containers, or large hollow containers. Although these same parts could be fabricated out of two separately formed sheets as illustrated later in Figure 11.1, the twin-sheet forming process can accomplish this in a much shorter time span. Basically, two processes are employed for twin-sheet forming. In a single-station machine, heating and forming of two sheets and the bonding process must be accomplished in sequence. With a four-station rotary thermoformer nearly 3 times the productivity can be obtained with no need of having to provide more than one pair of molds.

In one of the single-station machines, as illustrated in Figure 10.12 both plastic sheets are held in the same frame, just kept apart by injecting some air through a nozzle. In the oven both sheets are heated together from the outside only. Both sheets are simultaneously formed in the two opposing female molds by air pressure or vacuum. Because it is necessary to have both insides sufficiently hot for bonding these two parts together when the molds close, only fairly thin sheets can be employed.

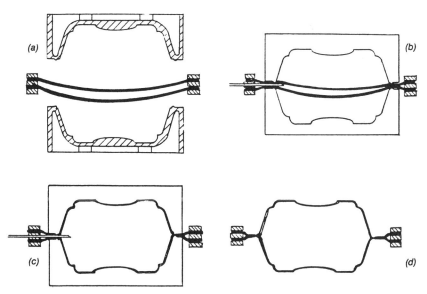

Figure 10.12. Twin sheet forming: (a) heated sheets inserted between mold halves, (b) molds closed and sheet edges heat bonded, (c) compressed air inflates part (air can also be injected through inserted needle), (d) manufactured part removed from mold.

When the single-station machine is provided with separate ovens on each side for each sheet, heat is supplied from top and bottom heater banks, thus obtaining better heat penetration also for thick sheets. The heated sheets are in short sequence shuttled to the center and each formed in the opposing molds, which are quickly clinched together.

The four-station rotary twin-sheet forming process requires a more sophisticated tuning in regard to heat input and timing. In station 1 two sheets are in relatively short sequence clamped in frame 1 and frame 2. These frames are heated in two ovens presenting stations 2 and 3. When the sheet in frame 1 reaches forming temperature, it is rotated to station 4 where it becomes formed in the lower mold, which must just temporarily be raised to accept the sheet. While the mold is lowered again, the frame opens up so that the empty frame can be rotated back to the first station. Simultaneously, the second sheet advances from the heating station 3 to the forming station 4. While the sheet is formed in the upper mold half of station 4, during the coming long time span the two sheets for the next part are heated. The lower mold is raised, and both halves are bonded together. Within the same time span the part must be cooled, and the first sheet for the third production part loaded. In the final sequence both molds are retracted, the frames rotated to station 1, the finished part removed and replaced with a new sheet for the advancement into the oven. Within the time of a full rotation two parts are being produced.

Because accuracy is important for twin-sheet forming the machine must be built sturdy. Figure 10.13 shows the detail of such a ring wheel assembly with the closed clamp frame at the left.

Important aspects in twin-sheet forming are:

(1) Exact vertical and perfect parallel alignment of both molds is imperative.
(2) Sufficient pressure must be applied to the pinch area to obtain a good bond. In case the part will be trimmed flush at the sides, the pinch area must be wide enough to exert an inward pressure for the formation of a pad at the inside. Raising the heat by additional heaters there will improve the bond and will become essential when forming high-temperature plastics. General pressure-forming techniques are, therefore, frequently brought into play.
(3) Because the cooling cycle will always determine output, cooling must be made as efficient as possible. Therefore, only aluminum molds should be considered. Internal cooling can be improved by inserting a number of blow pins, half of them for injecting cold air and another set distributed within the whole cavity to let hot air escape.
(4) The best area for the accommodation of inserts or openings when forming containers is along the pinch line.

Figure 10.13. Ring wheel assembly detail of a double oven, four-station twin-sheet pressure former (courtesy of MAAC Machinery Corporation, Itasca, IL 60143).

(5) In structural parts orphan pieces can be inserted just before the two sheets close up. They can consist of wood, solid, or foamed plastic but also of metal (threaded inserts) and serve as reinforcement or as substrate for assembly fasteners.

A process resembling twin-sheet forming is widely practiced in the thin film form-fill-seal technique for the efficient packaging of mostly small portions of food.

The twin-sheet forming process is gaining acceptance due to the possibility for producing large extended parts that excel in stiffness and low weight. In regard to labor intensity this process is positioned between the much slower rotational molding process using a less costly raw material and the more costly blow molding process, both of which must be considered to be stiff competitors.

Pressure forming

In recent years pressure-forming techniques have found widespread applications because they enabled the thermoformer to compete better with injection molders. Many parts can now be produced by this thermoforming process which, in regard to surface appearance, cannot be kept apart from injection-molded ones. This process is especially suited for large parts and has, therefore, found many applications for top-of-the line industrial, medical, and business machine housings or components.

The processes described up to this point utilize a vacuum or an air pressure limited to practically 14 psi. Although these pressures suffice to shape a heat-softened plastic sheet or film into rough contours, they cannot force the plastic into distinct engravings and low- or high-relief designs of a mold. However, many of these outlined forming methods can accomplish that if the relatively feeble vacuum force becomes replaced by air pressures up to 100 to 150 psi. The parts produced by such a method will show surface details on the mold side, which makes them indistinguishable from injection-molded parts. (This is illustrated in Figure 4.3, page 66). Some of the long fiber-reinforced thermoplastic sheets, as described in Chapter Nine, can only be formed by either having this additional pressure at the former's disposal or by matched-mold techniques.

It must be stressed that this change makes it necessary to secure molds that have considerably higher strength, better surface quality and details, as well as vent holes or slots that are smaller and more prudently located. Molds and pressure boxes need a solid foundation on the mold tables. The seals must be well designed and the clamps strong enough to prevent the escape of pressurized air. Only in vacuum forming these conditions are resolved by themselves. Furthermore, all safety precautions relating to pressure vessels have to be obeyed. Although all these changes increase expenses, mold costs should still amount only to one-quarter of that for injection molds.

Mechanical thermoforming

A number of the previously described forming methods incorporate mechanical forming steps applied concurrently with the action of an air pressure differential. Those were the bar assists and the various applications using a plug.

When using mechanical forming processes the plastic sheet or film is usually heated to a slightly lower temperature, because higher forces can be safely implemented. If forming takes place at temperatures below the glass transition temperature of the plastic, the process is not considered thermoforming (see Chapter Twelve).

Edge canting is one of the simplest mechanical thermoforming processes when working with relatively thin sheets. Only that line in the material where the bend will occur must be heated. Several shapes and forms can be obtained as long as they do not have compound curvatures. Figure 4.5 (page 67) shows one possibility for heating a narrow line of a plastic sheet preparatory to bending it in a narrow radius. Because no stretching of the material takes place, there is no reduction in sheet thickness. However, due to the localized heating, bending, and cooling, undesirable stresses may become incorporated into the formed part. Annealing the part is suggested to minimize distortions, stress cracking, and crazing.

When 1/2" or thicker sheets have to be edge formed, such as the highly filled sheets used as marble replacements, the whole sheet must slowly be brought up to temperature uniformly.

PLUG-AND-RING FORMING OR RIDGE FORMING

In this process the heated sheet is held in a frame or ring, while a plug or plate from one or both sides pushes and forms the sheet. The plug mold does not have to have a solid surface; it could just be a plate with welded-on protrusions or profiles. The forming will be restricted to the ridges of that skeleton and the sheet stretched in a straight but angular plane between the protrusions of both mold halves as shown in Figure 10.14. By slanting the ridges at an angle off the horizontal line, the plane surfaces will change to curved shapes. Decorative panels can easily be made with rather inexpensive forms consisting of a plywood box on which are mounted metal profile cutoffs (Figure 10.15). If a vacuum or pressure is utilized in place of a counter mold, the surfaces acquire convex shapes. Ridge forming is a fast forming process, in which the formed sheet can be cooled from both sides by blowing air against it, and which results in very reproducible shapes.

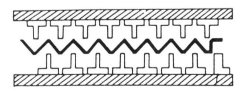

Figure 10.14. Ridge forms for corrugating plastic sheets (courtesy of Rohm and Haas Co., Philadelphia, PA 19105).

Figure 10.15. Ridge form for making decorative panels (courtesy of Rohm and Haas Co., Philadelphia, PA 19105).

SLIP FORMING

Occasionally, slip forming makes use of a forming technique common in sheet metal fabrication. The sheet, as seen in Figure 10.16, is not held tightly in a frame but is sustained solely by friction achieved with a predetermined spring pressure or a pressure exerted by pneumatic cylinders squeezing the sides of the forming frame. The frame should have a rounded edge and be made of a phenolic laminate or a temperature-resistant thermoplastic resin stock. During the forming process additional material from the circumference is admitted, to increase the draw depth. Cooling of the sheet in contact with the forming plate, greater sheet thickness requirements, and the risk of creases limit the usefulness

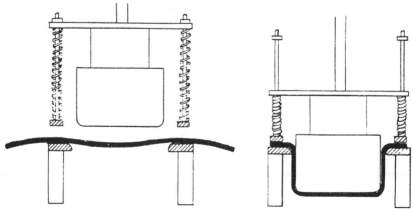

Figure 10.16. Slip forming, heated sheet is held only by spring-loaded bars.

of this process for thermoforming. Only materials that retain appreciable strength at forming temperatures are suitable for this process.

MATCHED-MOLD FORMING

A mechanical forming method that has developed into an important process is called *matched-mold forming*. It has found widespread applications in forming expanded polystyrene sheet for various kinds of packaging and display items. It is the ideal forming method for this type of material because it uses a lower temperature to avoid cell collapse, whereas the fluffiness of the heated sheet makes it possible to maintain forming pressure over the entire part despite any local mismatch between sheet wall thickness and mold clearance. The process for making egg cartons is sketched in Figure 10.17. If solid plastics are formed, it is important to provide sufficient cavity space to accommodate the thickest part of the sheet because the materials are too stiff to yield laterally within the mold cavity.

RUBBER PAD AND FLUID PRESSURE OR DIAPHRAGM FORMING

There are several alternatives in rubber pad and fluid pressure-forming methods. What they have in common is that only one mold half is required, that higher forming pressures are applicable, and—probably the greatest advantage—that all forming takes place under compression, which reduces the likelihood of tearing. The high mobility of low modulus rubber is utilized to act like a liquid and exert pressure onto all sides of a male mold. The heated sheet to be formed is placed between the rubber and the mold. For deeper parts, where a solid rubber pad would be too stiff, a rubber diaphragm is used, and the bulk of the rubber is replaced by a hydraulic liquid (Figure 10.18). The pressure of the liquid can be adjusted to expedite the forming process.

Other thermoforming processes

The preceding descriptions of forming methods focused on their main features. Modifications and customization of these processes are frequently developed. This is especially the case on fully automatic thermoforming machines, where production output is crucial. Several modifications were mentioned in the chapter covering packaging. Beginning with a material other than sheets or film is a further way in which thermoforming is modified. Use of lower priced starting materials, such as plasticized billets, requires specialized equipment. Experimental processes continue to enhance and replace established technologies.

The flow-forming process, sometimes also referred to as *melt flow stamping* utilizes fiber-reinforced thermoplastic sheets for the forming process. As described in Chapter Nine the fiber component can consist of glass, carbon, or aramid fibers. Although in many cases metal stamping equipment is being used, some of these materials are also shaped by regular thermoforming equipment based on vacuum or pressure forces. The sheet or a smaller sized billet must be heated above the melt temperature if the binder is a semicrystalline plastic, e.g., for nylon 6 approximately 525°F, and for polypropylene 400°F. If an amorphous

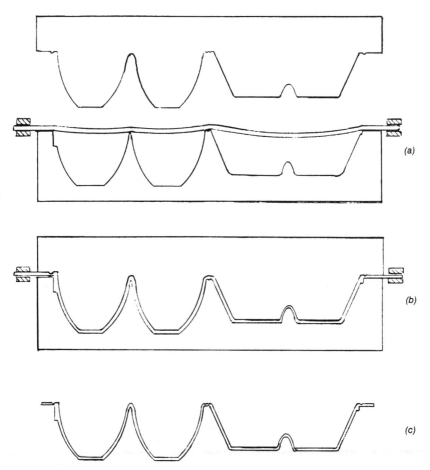

Figure 10.17. Matched-mold forming: (a) heated sheet held to female mold, (b) closed and part formed and cooled, (c) formed part.

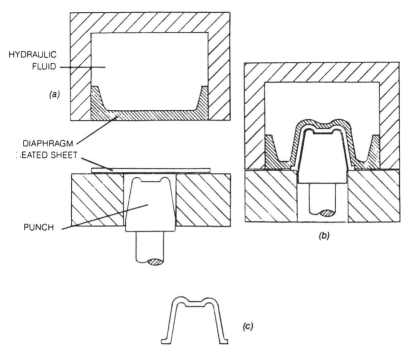

Figure 10.18. Rubber diaphragm forming: (a) heated sheet inserted in opened press, (b) press closed and punch raised while maintaining hydraulic pressure constant, (c) formed part.

resin is used as binder, the temperature must be higher than the glass transition temperature. The sheet is held in the press for up to 30 seconds at the maximum press stroke until it cools sufficiently. The (glass) fibers facilitate the handling of the molten plastic before forming and lend high rigidity to the part after forming.

Adjusting process parameters

When a part is produced for the first time or on a new machine, an expert who can diagnose the source of problems and make the required adjustments should be present. Individual expertise can be gained by having worked extensively either with thermoforming equipment or by experimenting with the same material and a simple laboratory thermoformer. Systematically varying the operating conditions and observing their effect on the quality of formed parts leads to a thorough understanding of particular processes.

The alteration of one condition usually affects other settings—if only to a limited extent. Accelerating the rate of production will require changing nearly all heater settings. The standby temperatures of molds, plugs, and shears may have no resemblance to the temperatures that prevail after, e.g., one-half hour of steady production.

One should always keep in mind that most control devices and meters indicate only a static condition (temperature, pressure, position, etc.). However, what must be mastered are the dynamic conditions, such as flow of heat (heating, cooling), flow of air (evacuation, pressurization), acceleration of molds or material, etc., to utilize the full capabilities of a process and of the machinery available. Newer process controllers furnish such information by scanning data points within very short time intervals and subsequently evaluate them for control purposes.

Difficulties after an extended production time are probably caused by malfunctioning of the equipment or a change in material. In general, thermoforming personnel are all-observant in recognizing aberrations in machine performance but less perceptive in regard to identifying material variations. Only problems involving appearance or gauge alterations can be clearly identified on the spot. Chemical or physical property aberrations will always require lengthy laboratory tests. The use of troubleshooting guides may occasionally be satisfactory. But given the plethora of materials and forming methods, it is not at all surprising to find two opposite states of affairs listed as probable causes for just one clearly definable problem.

Once acceptable parts have been produced consistently for a certain time, it becomes advisable to write down all the adjustable machine and process data. With today's proliferation of using computers both for process parameter control as well as production monitoring, it becomes very simple to put all present data in memory. This should also be done before it seems likely that more than one modification will have to be performed. Upcoming problems may usually be solved faster by first comparing all machine settings with the listed values and then checking whether all commands are executed properly. The following checks will help uncover burned-out heating elements, broken thermocouple wires, air leaks, plugged lines, motor malfunctioning, lubrication problems, etc. The questions to ask are:

(1) Is current flowing through all heaters as indicated?
(2) Is vacuum or pressure correct in the mold box?
(3) Is mold temperature correct (not to be confused with coolant temperature)?
(4) Are timing intervals correct?
(5) Are machine movements smooth?

To avoid getting caught with a surprise of having only nonsuitable material at hand, a new shipment or a new batch of material should be

PROBLEM	POSSIBLE CAUSES	SUGGESTED SOLUTIONS
Chill Marks	• Mold too cold. • Insufficient draft angle and radii.	• Mold should be heated (250-260°F). • Increase mold radii and draft angles.
Loss of Vacuum Seal	• Cold clamp frames. • Improper spacing between clamp frames and vacuum box.	• Preheat clamp frames (250-260°F). • Minimum space between clamps and vacuum box ½" to ¾".
Material Pulling out of Frames	• Insufficient clamp area. • Inadequate clamp pressure. • Uneven heating.	• Adjust clamp points uniformly at sheet perimeter. • Increase clamp pressure to maximum. • Control sheet temperature. Use center screening to allow more heat at sheet perimeter.
Excess Thinning Severe Necking Poor Surface Finish	• Drape speed too fast. • Improper forming temperature. • Mold design. • Incorrect forming technique.	• Drape speed should be set between 4-9 inches per sec. • Ideal forming temperature 375°F. • Increase upper radii. • Try snapback or billow-forming techniques.
Uneven Material Distribution	• Excess thickness variation. • Uneven heating. • Non-uniform clamp pressure. • Improper forming technique.	• Check gauge tolerances. • Check uniformity of heater output. Screen if necessary. • Maintain uniform clamp pressure to avoid pull-out. • Use billow or snapback forming method.
Wrinkles on Flat Horizontal Surfaces	• Uneven cooling due to slow drape speed. • Material too hot (too much sag or drape).	• Drape at higher speed 6-8 in/sec. • Screen center of sheet allowing edges to heat first. Use taller vacuum box to provide more pull by area.
Texture Washout and Excess Gloss	• Forming temperature too high. • Improper heating technique.	• Reduce heater inputs and cycle time. • Heat sheet from smooth side. Keep texture cool. • Precoat texture with strippable mask such as TR4997 by Spraylat Corp.
Pinholing or Pimples	• Vacuum holes too large. • Dust on mold or sheet. • Mold too cold/too smooth surface finish. • Vacuum rate too high.	• Use 50-mil holes or smaller. • Clean mold and sheet with deionizing airgun. • Keep mold temperature at 250°F. Draw sand mold surface with medium grit paper. • Place small orifice over main vacuum hole.

LEXAN sheet
Recommended Drying Times
at 260°F in a Circulating Oven

Thickness	Dry-Time
.093	3.5 hours
.125	4 hours
.187	12 hours
.250	24 hours

Minimum Mold Radius for LEXAN sheet
INSIDE RADIUS = 1 x T
INSIDE RADIUS = 1 x T

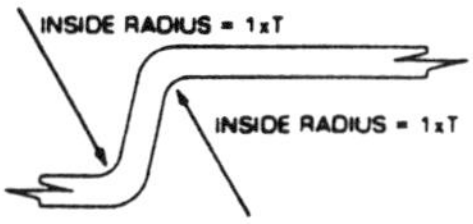

Minimum Clearance Between Vacuum Box and Clamp Frame
½"
MOLD
VACUUM BOX

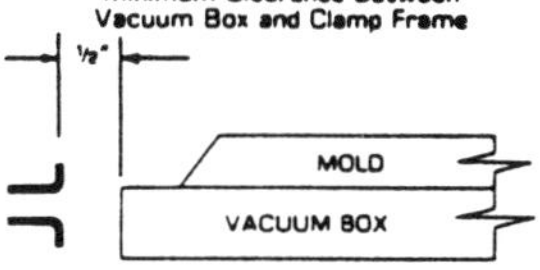

GENERAL ELECTRIC

MSC-430 (8/82)

Figure 10.19. The Thermoforming Troubleshooter (courtesy of General Electric Company).

LEXAN® sheet and film
THERMOFORMING TROUBLESHOOTER

PROBLEM	POSSIBLE CAUSES	SUGGESTED SOLUTIONS
Voids or Bubbles in Formed Parts	• Excessive moisture in sheet.	• Dry as recommended— 250°F for specified time with minimum separation of one inch between sheets.
Crazed or Brittle Parts	• Mold design. • Part left on mold too long. • The use of incompatible mold lubricants.	• Mold radii should be at least the thickness of material. • Remove part from mold as soon as it becomes form stable. • Use compatible powdered mold release.
Part Warpage	• Mold too cold. • Clamp frames too cold. • Part left on mold too long.	• Preheat mold 200-260°F. • Preheat clamp frames. • Remove part from mold as soon as it becomes form stable.
Non-Uniform Drape	• Uneven heating of sheet.	• Check heater section and adjust. Use selective screening if necessary. • Check for cold air drafts.
Difficult Part Removal	• Insufficient draft angle mold undercuts. • Mold finish perpendicular to direction of part removal. • Ejection pressure too low.	• Increase draft angle. Use strip rings or cam action mold. • Resurface mold. Sand mold sides vertically. • Add air holes, increase injection pressure. Use powdered mold release.
Poor Surface Finish	• Mold surface too rough. • Mold mark-off. • Draft angle.	• Draw-polish mold or use different mold material. • Use silicone or powdered mold lubricant sparingly. • Increase draft angle.
Insufficient Draw Down Poor Definition	• Improper sheet heating. • Insufficient vacuum. • Poor mold design.	• Increase heating time and temperature. • Check vacuum system for leakage. • Add more vacuum holes. Check for good seal between clamp frames and vacuum box.
Webbing or Bridging	• Improper mold layout. • Blank too large for mold. • Material overheated. • Improper mold design. • Vacuum rate too fast.	• Increase spacing between molds. Use grid or ring assist. • Leave minimum of material around mold. 2" is a good rule of thumb. • Shorten heat cycle. • Increase radii and draft angle. • Slow down vacuum rate (use smaller vacuum holes). Restrict main vacuum lines.

• Registered Trademark of General Electric Company   MORE ON REVERSE ▶

Figure 10.19 (continued). The Thermoforming Troubleshooter (courtesy of General Electric Company).

tried well before the satisfactory material runs out. Under no circumstances should the old material be used to the last sheet. It is only possible to pinpoint cropping up troubles on variations in material if samples from both the old and new material can be submitted to a laboratory. One should get into the habit of always retaining at least a dozen sheets of the old material. These can either be used—in case of trouble with the new material—to verify that the machine settings are all right or otherwise just be used up later. Laboratory tests conducted side by side with both old and new material will show differences in:

(1) Material dimensions, sheet gauge uniformity
(2) Surface imperfections
(3) Degree of orientation (affecting shrinkage and distortions)
(4) Impact resistance (falling dart impact test)
(5) Melt viscosity (melt index)
(6) Presence of humidity or other volatiles

After ascertaining that the above conditions are properly met, it is best to follow the material supplier's recommendations. Invariably, they will stress in their brochures which conditions are of importance and which must be followed. When process conditions are altered, it is wise to keep record of it. In a short time, one can compile one's own troubleshooting guide, which will lead to a very high percentage of correct assessments and successful readjustments.

Many of the difficulties and problems that may arise are reviewed in this book in those chapters that describe equipment, materials, and processing.

Thermoforming troubleshooting guide

Mainly for illustrative purposes (see pages 179–180), the Thermoforming Troubleshooter for Lexan Sheet and Film is reproduced in Figure 10.19. Again, it should be emphasized that other materials might require different adjustments to alleviate apparently similar problems.

ELEVEN
Design Considerations

DESIGN PERSPECTIVES HAVE been analyzed earlier in the context of specific processes and materials. A complete listing of all details is impossible in this book; therefore, only a few points will be sketched here.

Probably the first question to ask is whether there is a chance of producing the desired part by a thermoforming process, given its shape, size, wall thickness, etc. Then the questions of material properties, tooling, and processing should be evaluated. After these fundamental questions, all the details should be coordinated and specified. This process must be repeated several times, because the finalization of one aspect may unfavorably affect another. Details initially perceived as trivial may turn out to become stumbling blocks later.

(1) The designer must collect as much information as possible about that part's application, that is, what function it has to fulfil, how it fits in with the overall design when it is only one of the components, and what type of environment it will be exposed to.

Thermoformed parts are, by necessity, made of thermoplastics. If materials were ranked according to their expected longevity or perseverance to everyday environmental stresses, thermoplastics would not fare well. Such a listing with wide overlaps would read as follows: (a) thermoplastics, (b) plant or animal derived materials (paper, wood, leather), (c) thermoset plastics, (d) metals, and (e) ceramics and stones.

To define the environment is more difficult than it appears at first glance, since those conditions may change drastically at certain times of the part's use, e.g., during transportation, storage, inclement weather conditions, rough handling by inexperienced personnel, and the rest. The mechanical stresses form the primary concerns and include possible loading or pressures, flexing, impacting, and vibrational forces. Thermoplastics are quite sensitive toward changes in temperature and more or less aggressive chemicals. Combinations of these influences can prohibit their application or at least require the selection of a higher quality, more costly grade. Low-temperature exposure endangers brittleness, and excessive dimensional shrinkage and high temperatures, possible softening and buckling. Chemicals (including moisture) may lead to swelling, softening, polymer degradation, and if combined with mechanical tensile loading, to unexpected environmental stress cracks. Ultraviolet light can bring about polymer degradation and discoloration. The consequences of failure should not be overlooked and should give indications whether supply of replacement parts or other contingencies must be contemplated.

(2) Styling considerations should be looked at next. Besides functionability and ease of fabrication, the part must have a certain customer appeal, whether just an industrial part or a stylish consumer product. The required surface appearance, such as color coordination or matching, surface texture, and feel, should be considered.

Too often, an initially designed part with good appearance will lose its allure when later stiffening measures must be added, when odd hardware or means of fastening must be incorporated or when quality of trim edges cannot be maintained.

(3) Cost considerations can favor many times the selection of thermoformed parts. Thermoplastics represent mostly low-cost materials. Tooling costs can be low, allowing the production of low-volume parts or to fabricate parts for which market penetration remains questionable. Prototype or improvised tooling leads to very short lead times. This will also free up more time to do application testing, to improve design details, and explore labor cost reductions.

The quantity of parts required and the rate and duration of making them available are important cost concerns.

(4) The designed part must be concordant with thermoforming processes, thermoforming production–friendly. That entails all the subjects contained in this book. To refresh the high points, the following listing is given: selecting the lowest cost material that will satisfy the mechanical and appearance requirement at the given environment, selecting the best suitable mold with consideration for meeting dimensional tolerances, selecting a forming equipment and process that makes it possible to produce consistently parts of acceptable quality, including acceptable trim edges.

(5) Among the design considerations the customers' or the authorities' specifications or standards occupy an important aspect. Among the various government departments or trade organizations a few should be mentioned. Components that become incorporated in cars, trains, or airplanes must comply with strict flame spread rating and smoke generation standards (Department of Transportation, Federal Aviation Agency). Purity and toxicity concerns are regulated for all packaging and utensils that come in contact with food and drugs (Food and Drug Administration). Construction materials or components bear safety and durability requirements (Federal Home Administration, Veterans Ad-

ministration, Occupational Safety and Health Administration, and Underwriters' Laboratory mainly for electrical devices). On the basis of increasingly more generous jury tort awards it seems prudent to consider also possible legal implications and get liability insurance.

When the time comes to look for the best design details, all the experience collected over the years on the shaping of any plastic part remains relevant. The rules for taper or draft at the part sidewalls (and especially corners), rounding off edges, mold shrinkage, undercutting, and flash lines are applicable to thermoformed parts, too. In this case the latter would be trim lines. Others, such as bosses and reinforcing ribs, have no counterpart in thermoforming. By judiciously selecting a favorable shape, many headaches can later be avoided when tool design and production problems have to be solved. However, rules, although well established, are rules of thumb. Many parts have successfully been manufactured when a concession in one area led to the "transgression" of a rule in another. The chapter on thermoforming molds contains all the mold-related information, and the chapter on thermoforming processes, all the information needed for obtaining the desired wall thickness distribution.

Assembly and bonding

To fully utilize the capabilities of thermoformed parts, one should not limit the search just to autonomous parts that could possibly be produced by thermoforming. One must also expand one's view into areas where thermoformed parts constitute only a solitary part of a comprehensive assembly or where several of such formed parts must be combined.

The importance of dimensional accuracy and stability cannot be overemphasized for all such parts, and the reader is referred to the paragraphs on part shrinkage (pages 48–50). In many cases provisions for the assembly can easily be incorporated into the thermoformed part. Besides forming internal or external threads, nut plates can easily be encased by the formed sheet. Specific possibilities should be enumerated.

SNAP-FITS

Thermoformed parts are generally regarded as thin and flimsy objects; therefore, a snap-fit is in many cases an ideal assembly procedure. The necessary undercuts should be positioned away from outside corners, both for easy stripping from the molds as well as for facilitating assembly. In the locked-in position the thermoformed component gains in rigidity, and neither expensive hardware nor assembly time are required. One must ensure, however, that the plastic, once assembled, is not subject to intolerable stresses.

MECHANICAL BONDING

Riveting connections seem to be the most appropriate method for mechanical bonding. To avoid applying excessive force to the plastic part, the holes should be somewhat larger, and only low-strength rivets should be used (tubular rivets or aluminum pop rivets). To forestall failure due to environmental attacks, manufacturers of engineering resins recommend the application of silicone sealants to the holes prior to riveting. To prevent localized overloading, rivets should not be spread too far apart. Metallic washers may also be helpful to spread the forces caused by thermal expansion.

FORMING AROUND INSERTS

One very practical way of solving the assembly problem is the use of metallic inserts, which are placed on the mold so that their edges protrude, as in an undercut. These inserts can then serve as washers or threaded fasteners in the assembly.

WELDING

Excellent bond properties are always obtained with welding processes. Some of these processes are hot plate welding, hot air welding, heat sealing in the case of films, spin welding of circular parts, ultrasonic welding, microwave or electromagnetic welding (Emabond® System, Ashland Chemical Co., Columbus, OH, using ferrite-filled plastics), or electrical resistance welding when combined with metallic particles or wires. Ultrasonic welding equipment can also be used to provide a mechanical interference bond by deforming or staking certain areas of one of the thermoformed parts to be combined.

SOLVENT BONDING

Many thermoplastics can easily be bonded to themselves by solvent bonding. There are efficient ways for accomplishing this, if simple jigs are available. When a good fit is provided to the parts to be joined, the solvent will readily spread out by capillary attraction. Two cautionary remarks should be voiced. None of the useable solvents are indifferent to the plastic or the applicator:

(1) To obtain the full strength capabilities of the plastic, all traces of solvent must be removed by heating or annealing of the part. Because it is not always sufficient to let the solvent evaporate under ambient conditions, any detrimental effect on impact properties due to residual solvents should be monitored.
(2) Good ventilation or wearing of solvent vapor capturing respirators is necessary.

ADHESIVE BONDING

The great number of possibilities for adhesive bonding makes the selection of the right combination a formidable task. In most cases the plastics or the adhesive supplier will be able to reduce the number of trials considerably. Nonetheless, many factors can preclude the use of a promising adhesive. The simplest techniques, such as pressure-sensitive (foam) tapes or the use of hot-melt formulations, should always be tried first, before the stronger but slowly reacting single- or two-component adhesives, such as epoxies or acrylics, are explored. Some of the adhesive components, such as the above-mentioned solvents, may affect thermoplastic resins and result in the deterioration of impact properties.

Rigidizing thermoformed parts

The possibilities for improving the rigidity of the part by mold design or by the selection of materials (laminates) are covered in Chapters Three and Eight. Although these methods are relatively simple to apply, they are limited in regard to the improvements attainable. More effective rigidizing may be accomplished in certain cases where rigid strips, corners, or washers are positioned in the mold so that at least a part of the formed sheet can clasp them tightly. The intended design, however, must be probed to ensure that stresses from differential thermal expansion remain within tolerable limits. Three other methods that involve the use of several additional working steps and accessories are described here.

BONDING MULTIPLE PARTS

By bonding multiple thermoformed units together, very rigid struc-

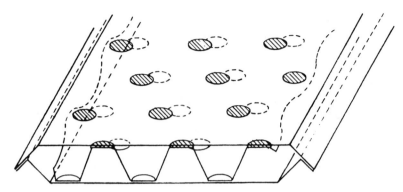

Figure 11.1. Laminated rigidized structure. Interlayer bonded alternately to upper and lower skin.

tural parts can be produced. The twin-sheet thermoforming process, which has gained much interest in recent years, represents an ideal way of accomplishing this. Figure 11.1 illustrates how an outer and inner thermoformed skin are bonded together along the outside edges. At the same time a somewhat smaller and thinner thermoformed sheet, which could have either deep corrugations or many formed-in dimples—alternately touching both outer skins at short intervals—will be bonded to both these skins with a structural adhesive. When only one smooth surface is required, the back side outer skin may be eliminated.

FOAMING-IN-PLACE

The foaming-in-place process could be regarded as a variation of the bonding process described above. Both thermoformed skins must be forcefully held together in a jig. Through an opening a predetermined amount of a liquid reactive polyurethane foam mixture is dispensed. The rising foam will bond to both surfaces and will be strong enough to allow removal of the finished part from the jig within a short time. Whether a rigid or semiflexible polyurethane foam system is used depends on the application. Figure 11.2 shows a thermoformed center car console rigidized with a semirigid polyurethane foam filling.

FIBER-REINFORCED STRUCTURAL SUPPORTS

Glass fiber-reinforced cured polyester resin parts have outstanding strength and rigidity. For such parts to also have a good appearance,

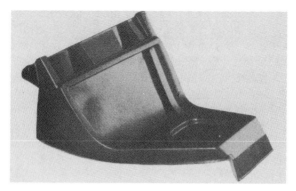

Figure 11.2. Center car console rigidized with semirigid polyurethane foam (courtesy of Adolf Illig Maschinenbau GmbH & Co., D-74080 Heilbronn, Germany).

excellent molds and the careful application of a gel coat are required. Savings can be realized, and the surface qualities of such large parts can be greatly improved by combining this manufacturing technique with thermoforming. From a single much lower cost mold a large number of thermoformed skins having not only the desirable shape but also an appealing surface texture and color can be rapidly produced. These skins in turn serve as molds for the glass fiber-reinforced unsaturated polyester resin lay-up or spray-up process, thus allowing the gel coat to be omitted. Because thermoformed skins might be rather flimsy, sturdy holding

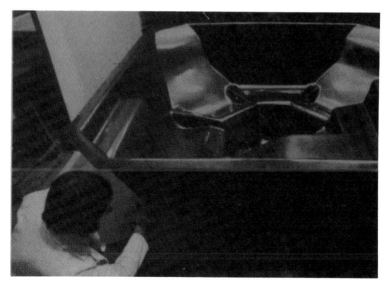

Figure 11.3. Inspection of vacuum-formed and rigidified spa (courtesy of Aristech Chemical Corp., Florence KY 41042).

fixtures must be provided for rolling out the air bubbles and maintaining the desired shape while the resin cures under exothermic conditions. Such fixtures may be constructed from wood or reinforced polyester resin without the need for laborious smoothing of the outer surfaces. A thermoplastic sheet that is compatible with the unsaturated polyester resin must be selected. The contact surface might have to be cleaned, roughened, or coated with an interlayer to ensure a lasting bond and prevent delamination during the extended use of such parts. Figure 11.3 shows a spa made according to this process.

Finishing and decorating thermoformed parts

As is the case for all manufactured products, appearance of any thermoformed part will greatly affect the customer's purchasing decision. The material's color and surface gloss are the primary features. But many variations of it can also be applied to thermoformed parts. There are possibilities to confer a frosted glass appearance to polyethylene terephthalate or styrene-acrylonitrile parts. As coloring agents, pearlescent or fluorescent dies can be added. Depending on the degree of colorant, mixing sheets with a marble look or a granite effect can be obtained. Embossing or printing can create textured surfaces or simulate cloth and wood patterns. Finishing and decorating steps are frequently applied to thermoplastics and are, therefore, also suitable for thermoformed parts. For general finishing procedures the reader is referred to other books and company publications.

Because printing of film is performed much faster than the printing of any formed part, preprinted films are frequently used for roll-fed automatic thermoformers. Many of these have the capability of scanning preprinted films to ensure correct registration of illustrations. To obtain well-proportioned pictures and inscriptions on diversely shaped parts, it is necessary to distortion print the flat film blanks.

Other decorating processes, such as hot stamping, flock coating, affixing of decals, and the application of antistatic or radio frequency interference coatings, should not be overlooked.

Care must be taken that high finishing temperatures or aggressive solvents do not impair the shape or the properties of the formed part.

If machining of formed parts, other than trimming, is required, they must lie on a flat surface equipped with protrusions for fastening them in a jig. This is especially important because thermoformed parts are usually not designed for high rigidity, unlike thermoset and glass fiber-reinforced injection-molded housings. Recommendations in regard to drilling, milling, or reaming procedures should be taken from the materials manufacturer's literature. In some cases annealing of the part to

Finishing and decorating thermoformed parts **191**

Figure 11.4. Thermoformed container which has provisions for properly stacking and nesting (courtesy of Portage Industries Corp., Portage, WI 53901).

relieve machining stresses may be advantageous (exposure to 10 to 20°F below the heat deflection temperature for 1 hour).

Attention must also be paid to the packaging and storing of thermoformed parts. To reduce required space, one may be tempted to stack formed parts tightly. This not only may cause deformations of the formed parts but could also lead to cracking during storage or shipping. Figure 11.4 illustrates on the left-hand side how incorporation of support steps in the design of the formed part can produce parts that can be placed securely on top of each other, whether for nesting empty containers or stacking filled ones.

TWELVE
Related and Competing Forming Processes

SINCE HEATING AND cooling of plastics are time-consuming processes and require a considerable expense in energy, thermoforming processes have been attempted at lower and lower temperatures. As is known, similar parts when stamped out of sheet metal require only a fraction of a second to be shaped. The lower the temperature, the stiffer plastics become and, concomitantly, the force which must be applied increases rapidly.

Unfortunately, within a certain temperature range, ductile plastic deformation changes, often quite suddenly, to a brittle fracture mode. This point marks the end of the usefulness of a particular forming process. At the boundaries of this change some tricks can be implemented, which have been discovered by studying the well-known brittle-ductile failure modes of the various thermoplastic materials.

The shape of the plastic dictates whether a ductile, lasting deformation, or a brittle fracture will occur. Fibers of plastics are the easiest to form. Naturally, they are confined to unidirectional deformation, just resulting in stretched filaments. Thin films list next and may be unidirectionally or biaxially stretched. If these processes were tried on thick sheets, brittle fracture would invariably occur beyond a limiting thickness due to the shear forces appearing in the perpendicular direction. The speed of deformation also affects the ductile-brittle transition. Slower motions usually succeed in preventing fracture. But, occasionally, a brisk rate of forming will prove beneficial, because internally generated heat under these circumstances can aid in preventing fracture. Most cracking phenomena are due to the appearance of tensile forces; therefore, under high isotropic ambient pressures the chance of brittle failure can mostly be prevented.

Upon heating during the thermoforming process the Van der Waal's forces between adjacent polymer chains are relaxed, thus letting the randomly coiled chains slide past each other so they can assume a more preferential location in the direction of pull. When such a formed plastic part is reheated, it reverts to its original shape, but to a limited extent and not forcefully. On the other hand, plastics oriented at lower than the glass transition temperature contain under force aligned and stretched polymer chain segments that will forcibly retract when reheated. Many film and tubing products utilize this phenomenon and have gained applications as shrink films and shrink tubings.

Because certain properties improve considerably with the application of any of these stretching methods, this type of orientation has found many uses. The properties most improved are modulus of elasticity and rigidity, resistance to fracture, and sometimes to environmental stress cracking. Chemical resistance is enhanced, but gas or vapor barrier properties only in some cases.

Forming processes performed at lower temperatures

A variety of names has been applied to low-temperature forming processes and products. The few practical applications now in use depend on selectively developed material combinations, very specific thicknesses, and specially detailed shapes.

In the solid-phase pressure-forming (SPPF) or nonmelt forming process, materials are heated not above the melting range but usually several degress below it. Although the sheet must still be heated, not much cooling time for the formed parts in the mold is required, because neither the retractive forces nor the gravitational force is high enough to cause deformation or warpage of the sufficiently stiff plastic part. This process, which can be licensed from Shell Development Company (Houston, TX), involves a polypropylene sheet that may contain an inner layer of a barrier type resin, e.g., ethylene vinyl alcohol copolymer or polyvinylidene chloride. The sheet is heated to just below its crystalline melting point (300 to 320°F for polypropylene). The lip or rim of the container is formed when the sheet is held in place above the tool cavity. The descending plug stretches and simultaneously orients the sides and bottom of the container. A blast of cold air shortens the cycle to 22 strokes per minute.

Another high-pressure, low-temperature forming process has been developed by the Petainer Development Company, a joint venture with the Sewell Plastics Company (Atlanta, GA).

Not much is yet known about a process in which containers are formed by rows of articulating blades. This Hitek CD (cuspation-dilation) forming process does not require a firm clamping of the sheet. The preformed parts can be vacuum formed to attain the final stage. (Although developed in Australia, it is licensed in the United States by Continental Can Co., Syosset, NY, and Sonoco Products Co., Hartsville, SC).

For the production of food containers, Dow developed their scrapless forming process (SFP), in which coextruded barrier material is first forged into round billets and then converted to the final shape through a plug-assisted pressure forming process (also called *COFO process*).

A variation of these processes was anounced in 1996 by QuesTech Packaging Inc. (Newport News, VA). In this scrapless melt-phase forming process all of the material cut out of (barrier-layer containing) sheet material is held in carrier rings, heated, and transferred into a multicavity mold. The near-melt phase, semisolid is formed under use of a plug-assist in a hydraulic forming station.

The designation warm forging is generally used when the above process is applied to parts with massive cross sections. The heating times may extend beyond 1 hour for 1-in.-thick blanks, which is not a problem if preheating ovens are available. In some cases microwave heating would be a possibility.

Cold forming or cold stamping processes should be operable at ambient room temperature. But in many cases preheating is suggested. Polycarbonate resin sheets are truly cold formable. This material can be subjected to all the forming processes widely used in metal fabrication without the need for any pretreatment or heating. In metallurgical terms the processes are stamping, cold drawing, cold heading, cold rolling, press brake bending, coining, explosive forming, spin forming (probably the toughest), and rubber pad or diaphragm forming (the easiest). Figure 12.1 shows a 19-in. diameter distribution transformer cover that was formed by the author in exactly the same way and at the same temperature (70°F) as a sheet metal blank of comparable thickness. The operations included blanking, punch drawing, prick-punching, and beading. Some special acrylonitrile-butadiene-styrene copolymers have been devel-

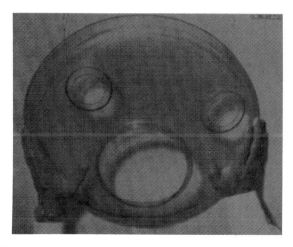

Figure 12.1. Polycarbonate distribution transformer cover formed on metalworking machines at ambient temperature (courtesy of *Modern Plastics,* September 1960, p. 137).

oped, which come close to the capabilities of polycarbonate. Cellulose butyrate resins, high molecular weight polypropylene, and the high density polyethylene resins have further reduced workability. That only a few formulations of plastics are suitable for these processes should not come as a surprise. In metallurgy too, alloy composition and crystal structure must be kept in narrow ranges to avoid fracture during the forming of more textured parts.

Still, the freedom one has with metal cold forming processes cannot be attained with plastic materials. A 1-in. copper rod can be reduced to a very fine wire by the use of successive cold drawing and annealing steps. In a plastic, consisting of randomly coiled polymer chains, structural disintegration eventually occurs when drawing or stretching is continued beyond the point where all chain segments are aligned (a few hundred percent). Given the suitability of a few plastics for metallurgical processes, specific precautions must be met. Both the dies and the sheet materials must be free of surface irregularities, and lubrication is essential. The elastic spring-back is considerably larger than that experienced with metals. Therefore, in many cases the extent of forming with plastics must be exaggerated. Sometimes, excessive frozen-in stresses will have to be relieved by solvent vapor treatment or by annealing.

Plastic materials that contain the aromatic (benzene) ring in their chain links will more readily withstand elevated temperatures. Cold-formed polycarbonate resin parts remain useful up to the boiling point of water with only negligible spring-back. The advantages of practicing these forming methods reside primarily in high-production rates, which equal those of metal forming and in the wide availability of appropriate machinery. However, due to much lower forming forces, as expected when one compares yield strength values, less powerful equipment may be used. Trim losses are considerably lower than in thermoforming because the material must not be held in a frame.

Packaging container forming

A large number of processes have been developed to satisfy the large market for packagings. Some may just remain pet projects; others may become obsolete in time. Only very few will dominate various markets over a long time.

Because sheet or film is mostly used in thermoforming as starting material, a few different approaches should be described here. A machine known as *Monaformer*, capable of producing 1200 parts per hour, goes through a number of individual processing stages. First, a reciprocating extruder converts the thermoplastic pellets into a melt, which is exactly metered into an open compression mold. When the mold is closed, a disc with a surrounding lip is formed. In the second stage the disc is held in a thermoforming device by its clamped lip. A plug stretches the preform disc toward the female cavity, and the material is conformed to the cavity by removing the air between the cavity and the preform on one side and by applying pressurized air to the other side of the preform. This forming force results in a certain biaxial orientation of the plastic, which contributes better mechanical properties to the formed containers. The outstanding improvements over conventional thermoforming are the compactness of the total equipment, savings in energy, and complete elimination of trim losses.

The Co-Pak process by National Can Co. (Jacksonville, FL) is also a multifaceted process. It starts with the thermoforming of a barrier laminate. Around this liner the structural plastic is injection molded and the thus formed composite is subsequently inflated to the size of the packaging container.

Another combination of operational steps is employed in the Vercon, Inc., container forming process. Two round container halves are first thermoformed. After trimming, the pairs are spin welded to obtain the 6-oz juice container for Campbell Soup Company.

Limitations for thermoforming

All the advantageous circumstances comprising thermoforming processes have been interpreted in the preceeding chapters. These advantages can be summarized by:

(1) Very low start-up costs due to (a) low costs for processing equipment and tooling based on low requirements for providing heating and forming pressure and (b) short lead times for getting into production
(2) Improved properties and more conveniences due to the use of higher molecular weight polymers with higher qualities and the possibilities of utilizing laminated, surface improved, textured, or already decorated film or sheets
(3) The practicability of producing parts that cannot be made any other way, oversized parts, exceptionally thin structures, or contoured foam sheet

There are two main disadvantages connected with the common thermoforming process:

(1) The starting material consists of a rather expensive intermediary component, an already processed film or sheet product. The granular

bulk materials that can be used for most other plastics molding or forming processes will invariably be of lower cost.

(2) Large areas outside the mold are needed for holding and transporting the sheet, so that only 20 to 85% of it will end up in the final product. Reprocessing of the trim material is more involved because it has to be first extruded again.

Another limitation for thermoforming is that close tolerances are more difficult to maintain. This becomes more pronounced when dimensional consistency becomes expected. This can only be achieved when the sheet or film material with undeviating measurements and properties is guaranteed and process conditions remain under tight control. Furthermore, mold details that are limited to one side only are more difficult to reproduce—with the exception of pressure-formed parts.

All of these disadvantages become more pronounced when production reaches high volumes.

To ensure that plastics parts are produced by using the most suitable method and the most economical material, the designer of thermoformed parts must also be familiar with the competing processes that are compiled in this chapter.

INJECTION AND EXTRUSION MOLDING

Injection and extrusion molding represent two of the competing processes that may be favored if the thermoforming designer cannot meet effectively all the requirements. Ribbing and variations in wall thickness contribute greatly to strength but cannot be obtained with thermoforming. Other applications have been reserved only for thermoforming. Examples are as follows: (1) food packaging where laminates and coextrusions are required either for barrier properties or for heat sealable top layers, (2) parts requiring ultraviolet resistance where laminates must be used for cost reasons, (3) parts bearing preprinted decorations or markings, and (4) parts for which it is desirable to frequently alter surface texture or color by using various embossing designs or pigmented sheet materials.

Compression and transfer molding, which are mainly employed for molding thermoset materials, are less encountered as competing processes. On the other hand, reaction injection molding and structural foam molding, processes that apply working pressures between that of pressure forming (approximately 100 psi) and injection molding (approximately 10,000 psi) might become used for the production of larger parts in competition with thermoformed parts. Such a related process for shaping fiber-reinforced thermoplastic sheets in a process called *thermoplastic composite sheet stamping* has been described in Chapter Nine.

BLOWMOLDING AND ROTOMOLDING

All three blowmolding processes (extrusion-, injection-, and stretch-blowmolding) are, in reality, thermoforming processes in which a cylindrically extruded or injection-molded tube serves as the starting material instead of a flat sheet. Certain indistinguishable hollow parts can be produced as well by the blowmolding process as the twin-sheet thermoforming process. One significant advantage of blowmolding is that during parison extrusion wall thickness variations can be programmed and different materials for different sections of the piece can be obtained either by coextrusion or sequential extrusions. On the other hand, the twin-sheet thermoforming process can accommodate two sheets of differeent color or thickness.

Although rotomolding and blowmolding processes were primarily developed for the production of containers that have only a very small fill hole but are still capable of holding a large volume, they are now used also for the production of parts originally produced by thermoforming processes. By cutting a rotomolded part into two pieces—not necessarily along only one plane—two identical trough-shaped parts or a container and its lid are easily fabricated. On the other side, bottles or hollow containers can be produced by joining two thermoformed top and bottom halves thermoformed from sheet stock. Bonding can be accomplished by pressure welding, by adhesives, or with the help of a spin-welding process.

PLASTISOL OR SLUSH MOLDING

The plastisol or slush molding process—mostly restricted to more or less plasticized polyvinyl chloride parts—represents another process that will also yield similarly shaped parts of uniform wall thickness. Although raw material costs and mold costs are low, long processing cycles make this process economically unattractive for the production of parts that can be shaped by alternatives, such as thermoforming.

Appendix A

Exemplary properties of thermoforming materials

THE PROPERTIES OF materials most frequently used in thermoforming processes are listed in two tables (Appendix A and Appendix B). As heavy gauge and thin film formed parts are differentiated throughout the book, a separation of materials has also been chosen for the two different processes. The table in Appendix A refers to the material properties prevailingly of interest for heavy gauge structural parts. Therein are listed the important mechanical properties and the thermal boundaries for their application. The specific gravity and the prices reflect 1997 quotations for general purpose; natural color extrusion compounds in truckload quantities are listed because material costs occupy the overwhelming portion of the cost of durable structural parts. The relative price helps to establish potential savings when substituting higher specific gravity materials with such of lower specific gravity. In some cases, sheet products can be obtained from primarily sheet producing companies at identical or lower prices than extrusion molding compounds.

Material	Flexural modulus, 10^3 psi, D790	Izod impact strength notched, ft lb/in., D256	Deflection temperature at 264 psi, °F, D648	Melting temperature, °F, crystalline, T_m
Properties ASTM Test Method				
Polymethyl methacrylate, PMMA	450	0.4	205	320
Polyacrylonitrile copolymer, PAN	550	1.5–4.5	164	275
Cellulose esters				
cellulose acetate, CA	190	0.5–5	110–190	445
cellulose acetate-butyrate, CAB	200	0.8–6	115–200	285
cellulose propionate, CAP	210	—	125–225	375
Polyolefins				
low-density polyethylene, LDPE	35–50	NB*	110 (66 psi)	210–240
high-density polyethylene, HDPE	145–225	0.4–4.0	180 (66 psi)	265–280
polypropylene, PP	150–300	0.4–1.5	120–140	320–345
polymethylpentene, PMP	100–190	2–3	125	445–465
ethylene vinyl acetate copolymer, EVA	8	NB*	115 (66 psi)	215–230
Styrene polymers				
polystyrene, PS	380–490	0.4	170–202	—
high-impact polystyrene, HIPS	160–390	1–7	170–205	—
styrene-acrylonitrile copolymer, SAN	500–600	0.5–0.6	200–220	—
acrylonitrile-butadiene-styrene cop., ABS	280–450	4–8	180–220	—
Halogen containing polymers				
polyvinyl chloride, rigid, PVC	300–500	0.5–20	140–170	415
polyvinyl chloride, plasticized	—	—	—	—
fluorinated ethylene-propylene cop., FEP	90	NB*	158 (66 psi)	525
Engineering plastics				
polyethylene terephthalates				
glycol copolymer, PETG, APET	300	1.9	147	—
crystallizable, CPET	350–450	0.3–0.7	70–150	420–500
polycarbonate, PC	340	16	270	455
polyamide 6, PA6	140	0.6–2	155–185	420
modified polyphenylene ether, PPE, PPO				
low glass transition	260	3–6	175–215	—
high glass transmission	300	5	225–300	—
polysulfone, PSU	390	1.1	345	—
polyetherimide, PEI	480	1.0	390	—
polyethersulfone, PES	370	1.5	380	—
polyether-etherketone, PEEK	560	1.6	320	635
polyamide-imide, PAI	730	2.7	532	—
polyphenylene sulfide, PPS	550	0.5	215–275	535–555

*No break

Melting temperature, °F, amorphous, T_g	Specific heat, Btu/lb °F, C351	Thermal conductivity, Btu ft/sq ft hr °F C177	Linear coefficient of thermal expansion, 10^{-5} in./in. °F, D696	Water absorption 24 hrs, %, D570	Specific gravity, g/cm^3, D792	Price, $/lb	Relative price, ¢/cu in.
220	0.35	0.108	3.5	0.2	1.19	1.05	4.6
203	0.4	0.15	3.7	0.28	1.15	1.08	4.3
160–250	0.3	0.12	10–15	2–7	1.3	1.56	7.3
248	0.4	0.12	11–17	1–2	1.2	1.54	6.7
—	0.4	0.12	11–17	1–3	1.2	1.54	6.7
—	0.55	0.2	10–22	<0.01	0.92–0.93	0.50	1.7
—	0.45	0.28	11	<0.01	0.95–0.96	0.45	1.5
—	0.8	0.1	8	0.01	0.90–0.91	0.35	1.2
—	0.91	0.1	7	0.01	0.83	—	—
—	—	0.2	9–11	0.01–0.1	0.92–0.94	0.70	2.5
165–220	0.32	0.09	4–8	0.01–0.03	1.05	0.44	1.6
200–220	0.34	0.05–0.15	5–10	0.05	1.06	0.46	1.7
220–390	0.25	0.07	6–7	0.2	1.08	0.86	3.3
190–250	0.3–0.4	0.11–0.19	3–7	0.2–0.4	1.02–1.08	0.90	3.4
165–220	0.25–0.35	0.1	5–10	0.01	1.50	0.35	1.9
—	0.3–0.5	0.1	4–14	0.15–0.7	1.3	—	—
—	0.28	0.14	5	<0.01	2.15	10.05	76
180–195	—	—	—	0.13	1.27	0.98	4.5
155–175	—	0.08	6.5	0.1–0.2	1.3–1.4	0.60	2.5
300	0.3	0.12	3.8	0.15	1.2	1.40	5.9
—	0.5	0.14	4.5	1.5	1.1	1.31	5.9
212–235	0.5	0.09	4–7	0.08	1.07	1.50	5.8
245–375	0.5	0.13	4–8	0.10	1.07	1.80	7.0
370–375	—	0.15	5	0.3	1.24	4.40	19.8
420	—	—	5.0–5.5	0.25	1.27	4.65	21.5
—	0.54	0.10	3.1	2	1.37	5.70	30
—	—	0.14	4.5	0.1	1.31	29.50	130
525	—	0.15	3.1	0.3	1.42	19.75	114
190	—	0.15	3–5	0.05	1.35	—	—

Refer to Appendix D (page 225) for the conversion factors for listed property values.

Appendix B

Exemplary properties of film materials

THE MATERIALS LISTED in Appendix B are preferably chosen for packaging applications. Since they have just a short-lived use, they are converted to thin films yielding either flexible containers, or if rigidity is required stiffening may be accomplished through surface structuring. Important properties are burst and tear strength. Especially for food, medical, and aerospace packaging, gas and moisture transmission rates must be considered. Because of great variations in film thickness, cost comparisons per weight are not very useful; therefore, their prices are no longer listed in this edition.

Material	Ultimate tensile strength, 10³ psi, D882	Elongation at break, % D882	Burst strength, 1 mil film, Mullen points, D774	Initial tear strength, lb/in. D1004	Propagation tear strength (Elmendorf), g/mil, D1922
Cellulosic films					
1. Cellulose acetate (CA)	7–14	15–50	40–80	1–2	3–12
2. Cellulose acetate butyrate (CAB)	5–9	50–100	40–70	80–100	5–10
3. Regenerated cellulose, coated	8–18	15–40	20–40	110–510	2–15
Polyolefin films					
4. Polyethylene, low density (LDPE)	15–3	100–700	10–40	65–575	100–300
5. Polyethylene, high density (HDPE)	2.5–5	25–600	N.A.	N.A.	20–200
6. Polymethylpentene (PMP)	2.5–4	10	—	400–650	—
7. Ethylene vinyl acetate copolymer (EVA)	1–3.5	500–800	10–15	80–500	50–200
8. Ionomer	3–4	300–400	—	—	25–100
Halogen atoms containing films					
9. Polyvinyl chloride, rigid (PVC)	7–10	25–50	30–40	110–490	10–700
10. Polyvinyl chloride, plasticized	2–5	100–400	20	110–290	60–1000
11. Polyvinyl fluoride (PVF)	8–11	95–110	17–46	330–500	17–46
12. Fluorinated ethylene propylene (FEP)	3	300	11	600	125
13. Polychlorotrifluoroethylene (PCTFE)	5–10	50–150	23–30	330–900	2.5–40
14. Ethylene chlorotrifluoroethylene cop. (ECTFE)	8–10	150–250	35–40	350–500	900–1300
Biaxially oriented films					
15. Polypropylene (OPP)	7–24	30–300	—	1000–1500	5–10
16. Polystyrene (OPS)	8–12	3–30	13–35	270–500	5–25
17. Polymethyl methacrylate (PMMA)	8–12	5–20	—	340–380	—
8. Poly(ethylene terephthalate) (OPET)	20–30	70–130	50–75	1000–3000	50–200
Polyvinyl chloride (OPVC)	8–16	70–350	—	—	30–400
Vinylidene chloride copolymer (PVdC)	7–20	20–40	30–60	2	10–30
olyamide (nylon 6)	30	50–120	140	300–500	16–28
mportant films					
olyester (PETG)	8.6	400	—	—	140–170
arbonate (PC)	9	85–105	35–40	1100–1600	45–55
mide (nylon 11)	8–14	250–400	190	1000–1100	400–500

endix D (page 225) for listed property values.

Change in dimension at 212°F (30 minutes), % D1204	Gas transmission rate at 1 atm and at R.T $\frac{cm^3 \cdot mil}{100\ in^2 \cdot 24\ hr \cdot atm}$ D1434			Moisture vapor transmission rate at 100°F $\frac{g \cdot mil}{100\ in^2 \cdot 24\ hr}$ E96(E)	Area factor, $\frac{sq\ ft \cdot mil}{lb}$
	Oxygen	Nitrogen	Carbon dioxide		
+0.2 to –3	100–150	30	1000	20–30	145–150
+0.1 to –3	600–1000	200–300	6000–10,000	30–40	160
–0.7 to –3	0.6	1	0.5–5	1–5	140–160
–2	500	180	2000–2500	1	210
–0.7 to –3	150	30	500	0.5	205
—	—	—	—	—	230
<1	850	400	3000	3	205
—	300–400	50–100	500–1000	1.5–2	205
–7 to +4	15–20	10–20	30–60	2–5	140–160
± 10	20–750	10–60	75–3000	3–15	135–150
–5 (338°F)	3.2	0.25	11.1	957	140
<1	750	320	1650	0.4	90
± 10	10–15	2	15–30	0.03	90
–3 to +2.5 (300°F)	25	10	110	0.6	115
< –1 to –20 (2 min)	150	30–50	500	0.5	215
–5 (2 min)	300–350	50	900–1400	5–10	185
< –1 (2 min)	—	—	—	1	160
< 1 to –40 (2 min)	3–5	1	20	2	140–160
–5 to –40 (2 min)	20–200	3–30	80–800	3–30	130–150
–3 to –40 (2 min)	0.1–1	0.1–1	1	0.1	115
<–2.5	2	1	10	10–20	165
—	25	—	125	4	160
0	190	50	800	6.4	160
<–3	30–60	5–15	150–300	3–5	190

Appendix C

Trade names and materials manufacturers

Listing of trade name materials.

Trade name	Manufacturer	Chemical identification	Heat deflection temperature (264 psi) or application
Acclear	Amoco Polymers	High-clarity propylene random copolymer	Microwavable, 208–255°F (66 psi)
Accpro	Amoco Polymers	Enhanced propylene polymer	Freezer containers, 190°F (66 psi)
Acctuf	Amoco Polymers	Impact propylene copolymer	High moisture barrier film
Aclar	AlliedSignal	Polychlorotrifluoroethylene film	165–212°F
Acrylite	Cyro Industries	Acrylic extrusion compound	181–189°F
Acrylite Plus	Cyro Industries	Impact modified acrylic extrusion compound	170°F
Acrysteel	Aristech Chemical Corp.	Continuous-cast impact resist. acrylic sheet	
Acrystone	Aristech Chemical Corp.	Continuous-cast mineral filled acrylic sheet	
Adcote	Morton International	EVA and/or acrylic based coating formulation	Heat seal and/or overprint coating
Adheflon	Elf Atochem	Polyvinylidene fluoride resin	Tie-layer resin
Admer	Mitsui Petrochemicals	Functional groups containing polyolefin	Coextrusion tie-layer
Affinity	Dow Chemical	Polyolefin plastomers	Superior packaging, sealants
Ageless	Mitsubishi Gas	Oxygen absorber for films	Fresh food packaging
AIM	Dow Chemical	Advanced styrenic resins	179°F
Akuloy	DSM Engineering Plastics	Polypropylene modified nylon alloy	150°F
Altair	Aristech Chemical Corp.	Opaque superstrong acrylic/thermopl. laminate	
Amodel	Amoco Perf. Prod.	Semicrystalline polyphthalamide	248°F
Amosorb	Amoco Chemicals	Oxygen scavenger concentrate	Food packaging
Apec	Bayer Corp.	High-heat polycarbonate	284°F
Appeel	DuPont Packag.	Coextrudable coating resin	Peelable lidding sealant
Arcel	Nova Chemicals Inc.	Polyethylene copolymer foam resin	Packaging
Ardel	Amoco Polymers	Aromatic polyester, polyarylate	
Attane	Dow Chemical	Ultra low density linear ethylene copolymers	Clear packaging
Azdel	Azdel, Inc.	Glass fiber mat reinforced polypropylene sheet	310°F
Azloy	Azdel, Inc.	Glass fiber mat reinf. amorph. engineering plastic sheet	290°F

210

Trade name	Manufacturer	Chemical identification	Heat deflection temperature (264 psi) or application
Azmet	Azdel, Inc.	Glass fiber mat reinf. cryst. engineering plastic sheet	425°F
Bapolan	Bamberger	Polystyrene and styrene copolymers and alloys	
Bapolene	Bamberger	Polyethylene and polypropylene polymers	
Bapolon	Bamberger	Polyamide 6 and 66 resins	
Barex	BP Chemicals Inc.	Acrylonitrile-methyl acrylate copolymer	High barrier polymer
Bayblend	Bayer Corp.	Polycarbonate/acrylonitrile-butadiene-styrene alloy	240°F
Bexloy W	DuPont Automotive	Modified ionomer resin	
Bicor	Mobil Chemical	Oriented polypropylene film	(Food) packaging
Biomax	DuPont Films	Biodegradable modified polyethylene terephthalate resin	
Bynel	DuPont Packaging	Ethylene-vinyl acetate terpolymer	Coextrusion adhesive tie-layer
Cadon	Bayer Corp.	Styrene-maleic anhydride terpolymer	205–250°F
Calibre	Dow Chemical	Polycarbonate resin	270°F
Capran	AlliedSignal	Polyamide 6 film (PVdC-coated)	Food packaging film
Capran Emblem	AlliedSignal	Biaxially oriented polyamide 6 film (PVdC-coated)	Food packaging film
Capron	AlliedSignal	Polyamide 6 resin and polyamide-ethylene copolymer	147°F
Carilon	Shell Chemical	Aliphatic polyketone resin	Engineering plastic
Celstran	Polymer Composites	Long strand glass fiber reinf. thermoplastic	
Centrex	Bayer Corp.	ASA and acrylonitrile-ethylene-styrene	Weatherable polymers
Cleartuf	Shell Chemical	Nucleated crystalline polyethylene terephthalate resin	400°F
Clysar	DuPont Co.	Biaxially oriented (irradiated) polyethylene shrink film	Packaging
Corian	DuPont Corian Products		Thermoplastic sheet
Corterra	Shell Chemical	Polytrimethylene terephthalate resin	
Cotie	BP Chemicals	Coextrudable tie-layer for polyacrylonitrile	Barrier packaging
Cycolac	GE Plastics	Acrylonitrile-butadiene-styrene terpolymer resin	200–215°F
Cycoloy	GE Plastics	Polycarbonate/acrylonitrile-butadiene-styrene blend	220°F

(continued)

Trade name	Manufacturer	Chemical identification	Heat deflection temperature (264 psi) or application
CyRex	Cyro Industries	Acrylic-polycarbonate alloy	214°F
Cyrolite	Cyro Industries	Acrylic-based multipolymer compound	163–186°F
Cyrolite HP	Cyro Industries	Acrylic-based multipolymer sheet	Interior signage, pop displays, 195°F
Cyrolon	Cyro Industries	Polycarbonate sheet	
Daran	Hampshire Chemical	Polyvinylidene chloride emulsion	Barrier coating
Daratak	Hampshire Chemical	Polyvinylacetate emulsion	Packaging film adhesive
DKE	Uniroyal Technology Corp.	Polyvinyl chloride/acrylic alloy	180°F
Delrin	DuPont Engg. Polymers	Polyacetal resin	255°F, 330°F
Dowlex	Dow Chemical	Linear low density polyethylene resin	
Durethan	Bayer Corp.	Polyamide 6	140°F/356°F (66 psi)
Dylark	Nova Chemicals Inc.	Styrene-maleic anhydride copolymer and alloy	225°F
Dylene	Nova Chemicals Inc.	Polystyrene moldable foam resins	
Dylite	ARCO Chemical Co.	Expandable polystyrene resin	Foam packaging
Eastalloy	Eastman Chemical	Polyester/polycarbonate resin	194°F
Eastapak	Eastman Chemical	Nucleated polyethylene terephthalate resin	Ovenable food trays
Eastar	Eastman Chemical	Cyclohexylenedimethylene terephthalate/isophthalate poly.	
Elite	Dow Chemical	High strength linear low density polyethylene	Packaging film resin
Elvax	DuPont Packaging	Ethylene-vinyl acetate (7–33 wt%) copolymer	(Food) packaging
Emaweld	Emabond System	Ferromagnetic compound	Thermoplastics assembly
Enduran	GE Plastics	Highly mineral filled polybutylene terephthalate	320°F
Epolene	Eastman Chemical	Low molecular weight polyolefin resins	Packaging adhesive
Escorene	Exxon Chemical	Polyethylene and polypropylene	
Ethafoam XL	Dow Chemical	Cross-linked polyethylene foam	
Eval	EVAL Comp. of America	Ethylene-vinyl acetate-vinyl alcohol copolymer resin	High barrier film
Everflex	Hampshire Chemical	Vinyl copolymer emulsions	Packaging adhesive
Expan	Chisso America	High-melt-strength polypropylene	

Trade name	Manufacturer	Chemical identification	Heat deflection temperature (264 psi) or application
Flair	AtoHaas North America	Patterned plastic sheet	
Flamtard	Twinpak Inc.	Fire-retardant polyolefin sheet	
Fluorosint	Polymer Corp.	Polytetrafluoroethylene based composition	Tooling
Fome-Cor	Solutia	Extruded polystyrene foam board	
Forex EPC	Alucobond Technologies	(Foamed) polycarbonate sheet	
Formion	Schulman	Ionomer resins	
Fortiflex	Solvay Polymers	High density polyethylene resin	
Fortilene	Solvay Polymers	Polypropylene resin	
Fortron	Ticona Corp.	Linear polyphenylene sulfide	518°F
Fusabond	DuPont Packaging	Maleic anhydride modified polyolefin resin	
Gapex	Ferro Corp.	Engineered polypropylene resin	Sheet extrusion
Gar-Dur	Garland Mfg.	Ultra-high-molecular-weight polyethylene	Tooling
Geloy	GE Plastics	Acrylonitrile-styrene-acrylate (PVC blend for capstock)	160–180°F
Geon	Geon Co.	Polyvinyl chloride compounds	
Grilamid	EMS-American Grilon	Nylon-12 and transparent nylons	
Grilbond	EMS-American Grilon	Adhesion promoter	
Grilon	EMS-American Grilon	Nylon-6, nylon-66 and copolymers	
Grivory	EMS-American Grilon	Partially aromatic nylon	
Halar	Ausimont USA	Ethylene-chlorotrifluoroethylene copolymer	High barrier material, 150°F
HiFax	Montell North America	Polyolefin resins	
HiGlass	Montell North America	Glass-reinforced polyolefin resins	
Hivalloy W	Montell North America	Polypropylene/acrylic alloys	130–145°F, weatherable
Hycar	BFGoodrich	Polyvinyl chloride/nitrile rubber alloy	
Hyflon MFA	Ausimont USA	Tetrafluoroethylene-perfluoromethylvinylether copolymer	Use up to 450°F
Hygard	Sheffield Plastics	Laminated polycarbonate sheet	
Hylar	Ausimont USA	Polyvinylidene fluoride resin	230°F, Chem. resistance

(continued)

Trade name	Manufacturer	Chemical identification	Heat deflection temperature (264 psi) or application
Hyrizon	Aristech Chemical Corp.	Sprayable rigidizing urethane foam	192°F
Hyzod	Sheffield Plastics	Polycarbonate sheet	257°F
Impax	GE Plastics	Opaque polycarbonate sheet	160°F, (Food) packaging
Implex	AtoHaas North America	Impact acrylic sheet	
K-Resin	Phillips Chemical	Styrene-butadiene block copolymer	
Kadel	Amoco Perf. Prod.	Semicrystalline polyketone resin	
Kaladar	DuPont Films	Polyethylene naphthalate ester resin	High temperature barrier film
Kama	Kama Corp.	Biaxially oriented polystyrene film	(Food) packaging
Kapton	DuPont Films	Aromatic polyimide film	High temperature film
Keldax	DuPont Automotive	Ethylene interpolymer, highly filled	Sound barrier sheet
Korad	Polymer Extruded Prod.	Acrylic clear/colored surfacing film	UV screen
Kraton	Shell Chemical	Elastomeric polystyrene block copolymers	(Food) packaging
Krystaltite	AlliedSignal	Polyvinyl chloride based film	Shrink packaging
Kydex	Kleerdex Co.	Polyvinyl chloride/acrylic alloy (fire-rated)	165–175°F
Kynar	Elf Atochem North America	Polyvinylidene fluoride (co)polymer	183–244°F
Lamal	Morton International	Reactive urethane prepolymer	Coextrusion adhesive
Lamisan	Charlotte Chemical	Calendered polyvinyl chloride sheet	
Lamicon	Toyo Seikan	Ethylene-vinyl alcohol copolymer	High barrier container
Lennite	Westlake Plastics	Ultra-high-molecular-weight polyethylene	Tooling
Lexan	GE Plastics	Polycarbonate resin and sheet	270°F
Lucite	ICI Acrylics	Cast acrylic sheet	203°F
Lumirror	Mitsui Plastics	Biaxially oriented polyester film	
Luran	BASF	Styrene-acrylonitrile copolymer resin	215°F
Luran S	BASF	Acrylonitrile-styrene-acrylate copolymer resin	
Lustran	Bayer Corp.	ABS terpolymer, SAN copolymer, PVC alloys	189°F
Magnum	Dow Chemical	Acrylonitrile-butadiene-styrene terpolymers	165–197°F

Trade name	Manufacturer	Chemical identification	Heat deflection temperature (264 psi) or application
Makroblend	Bayer Corp.	Polycarbonate/polyester alloy	190–248°F
Makrofol	Bayer Corp.	Polycarbonate film	
Makrolon	Bayer Corp.	Polycarbonate resin	270°F
Malon	M.A. Industries	Polyethylene terephthalate film resin	
Marlex	Phillips Chemical	High density polyethylene resin, polypropylene resin	129°F
Melinar	DuPont Films	Polyethylene terephthalate film	(Food) packaging film
Melinex	DuPont Films	Polyethylene terephthalate film resin	(Food) packaging film
Metapor	Segen & Co.	Porous, machinable blocks (Al-powder filled epoxy)	Mold material
Mindel A	Amoco Performance Prod.	Polysulfone/acrylonitrile-butadiene-styrene alloy	303°F
Mindel B	Amoco Performance Prod.	Polysulfone/polyethylene terephthalate alloy	
Modic	Mitsubishi Gas	Polyolefin, anhydride grafted	Coextrusion adhesive
Monax	Tredegar Film Products	Compression-rolled monodir. oriented HDPE film	Moisture barrier film
Moplen	Montell North America	Polypropylene	
Mxsten	Eastman Chemical	Extra-low density polyethylene resin	Film resin
Mylar	DuPont Films	Polyethylene terephthalate film	(Food) packaging
Mylar HS	DuPont Films	Polyethylene terephthalate film	Heat shrink film
Mylar M	DuPont Films	Vinylidene chloride copolymer coated PET film	High barrier film
Mylar OL or RL	DuPont Films	Polyethylene terephthalate film	Ovenable lidding
MXD6	Mitsubishi Gas	m-Xylylene diamine-adipic acid nylon resin	High barrier (co)extrusion resin
Neopolen	BASF	Moldable polyethylene foam	
Noryl EN	GE Plastics	Modified polyphenylene oxide	180–260°F
Noryl GTX	GE Plastics	Polyphenylene oxide/nylon alloy	355°F (66 psi)
Novodur	Bayer Corp.	Acrylonitrile-butadiene-styrene terpolymer	192°F
Novolen	BASF	Polypropylene resin	
Nucrel	DuPont Packag.	Ethylene-methacrylic acid copolymer resin	Foil and nylon adhesion
Nylopak	Dow Chemical	Coextruded polyolefin-polyamide film	(Food) packaging

(continued)

215

Trade name	Manufacturer	Chemical identification	Heat deflection temperature (264 psi) or application
Oppalyte	Mobil Chemical	Oriented polypropylene film laminate	Packaging
Optum	Ferro Corp.	Clay/calcium carbonate filled polyolefin alloy	190–240°F (66 psi)
Oxyguard	Toyo Seikan	Oxygen-absorbing plastic film layer	Oxygen barrier
OxyShield	AlliedSignal	Nylon 6 laminated ethylene-vinyl alcohol copolymer film	High barrier
Pebax	Elf Atochem North America	Polyether-polyamide block polymer	125–210°F (66 psi)
Pelaspan-Pac	Dow Plastics	Expanded polystyrene	
Pellethane	Dow Plastics	Polyurethane elastomers	
Pentafood	Klockner Pentaplast	Food grade film and sheets	
Pentaform	Klockner Pentaplast	Thermoform film and sheets	
Pentamed	Klockner Pentaplast	Medical grade film and sheets	
Pentapharm	Klockner Pentaplast	Pharmaceutical film and sheets	
Perspex	ICI Acrylics	Acrylic polymer resin and sheet	
Petite II	Shell Chemical	Crystallizable polyethylene terephthalate	Low-density cellular sheet
Petrothene	Millennium Petrochemicals	Polyethylene, high and low density and polypropylene	
Plexar	Millennium Petrochemicals	Modified polyolefin	Coextrusion tie-layer resin
Plexiglas	AtoHaas	Polymethyl methacrylate sheet	205°F
Pliovic	Goodyear	Polyvinyl chloride resin	
Poly 76, 84	Polycast Technology	Cast acrylic sheet	
Polyfabs	Schulman	Acrylonitrile-butadiene-styrene terpolymer compounds	
Polyflex	Plastic Suppliers	Oriented and biaxially oriented polystyrene film	
Polyman	Schulman	Acrylonitrile-butadiene-styrene terpolymer alloys	
Polyvin	Schulman	Polyvinyl chloride compounds	
Prevail	Dow Chemical	Polyurethane/ABS blend	165°F
Primacor	Dow Chemical	Extrusion compound	Heat seal or tie-layer
Pro-Fax	Montell North America	Polypropylene resins	
Pulse	Dow Chemical	Polycarbonate/acrylonitrile-butadiene-styrene alloy	238°F

Trade name	Manufacturer	Chemical identification	Heat deflection temperature (264 psi) mor application
Pyropel	Albany International	Structural thermal insulation	up to 550°F
QLF	BOC Coating Technology	Plasma enhanced chemical vapor deposition	Barrier coating
Quadrax	Quadrax Corp.	Carbon, glass or aramid fiber reinforced prepregs	130–250°F
Quarite	Aristech Chemical	Acrylic sheet with a granite appearance	
Questra	Dow Chemical	Semicrystalline syndiotactic polystyrene	210°F
Radel A	Amoco Performance Prod.	Polyethersulfone resin	399°F
Radel R	Amoco Performance Prod.	Polyphenylsulfone resin	405°F
Regaltech	O'Sullivan	Medical grade PVC and PP sheet and film	
Repete	Shell Chemical	Polyethylene terephthalate resin	
Resinite	Borden Packaging	Flexible food packaging polyvinyl chloride film	Food packaging
Retain	Dow Chemical	Polycarbonate/acrylonitrile-butadiene-styrene alloy	
Rexene	Rexene Products	Polypropylene and copolymer resins	up to 275°F (66 psi)
Reynolon	Reynolds Metals	Polyvinyl chloride shrink film	(Food) packaging
Rilsan	Elf Atochem	Polyamide 11 and 12 resins	275°F (66 psi)
Rovel	Dow Chemical	Styrene-acrylonitrile-olefinic elastomer alloy	210°F
Royalex	Uniroyal Technology Corp.	5-layer closed-cell ABS or PVC/ABS sheet	Canoes
Royalite R	Uniroyal Technology Corp.	Acrylonitrile-butadiene-styrene sheet	225°F
Royalstat	Uniroyal Technology Corp.	ABS/PVC or HDPE static control sheet	174°F or 156°F (66 psi)
Rynite	DuPont Engg. Polymers	Polyethylene terephthalate resin and elastomer alloy	444°F
Ryton	Phillips Chemical	Polyphenylene sulfide resin	500°F
Santoprene	Advanced Elastomer	Polypropylene elastomer alloy	
Saran	Dow Plastics	Polyvinylidene chloride film resin	High barrier food packaging
Saranex	Dow Plastics	Coextruded polyolefin-polyvinylidene chloride film	Food packaging
Selar PA	DuPont Packaging	Amorphous nylon resin as additive to EVAL resin	Improved coext., high barrier layer
Selar PT	DuPont Packaging	Modified amorphous ethylene terephthalate copolyester	Clear monolayer, high temp. barrier
Selar RB	DuPont Packaging	Laminar blend of polyolefin and nylon or EVOH resins	High barrier concentrate for packag.

(continued)

Trade name	Manufacturer	Chemical identification	Heat deflection temperature (264 psi) or application
Sintra	Alucobond Techn.	Rigid closed cell polyvinyl chloride foam sheet	High barrier coextrusion film
Soarnol	Morton International	Ethylene-vinyl alcohol copolymer resin	
Solacryl	Polycast Technology	Cast acrylic sheet	
Solef	Solvay Polymers	Polyvinylidene fluoride resin	302°F, (food) packaging film
Solmed	American Mirrex	Extruded polypropylene	Pharmceutical packaging film
Soltan	Solvay Polymer	Polyacrylonitrile copolymer resin	(Food) barrier packaging
Spectar	Eastman Chemical	Copolyester resins PETG	High clarity, 157°F
Spectrum	Uniroyal Technology Corp.	Polyolefin sheet	170°F (66 psi)
Stanyl	DSM Engineering Plastics	Nylon-4/6 resin	320°F
Stapron C, E	DSM Engineering Plastics	Polycarbonate/ABS or polyester alloys	
Stapron S	DSM Engineering Plastics	Elastomer modified styrene maleicanhydride resin	234°F
Stat-Rite	BFGoodrich Specialty	Static dissipative polyester alloys	
Styrofoam	Dow Plastics	Extruded polystyrene foam	Packaging
Styrolux	BASF	Clear styrene-butadiene block copolymer	
Styron	Dow Chemical	(High-impact) polystyrene	(182°F), 203°F
Sullvac	O'Sullivan	Polyvinyl chloride/ABS alloy sheet	
Supec	GE Plastics	Polyphenylene sulfide compounds	
Sur-Flex	Flex-O-Glass	Ionomer film	
Surlyn	DuPont Packaging	Ethylene-methacrylic acid salt copolymer, ionomer resin	Flexible film and skin packaging
Syntac	Emerson & Cuming	Glass sphere syntactic foam	Mold plug material, up to 450°F
Tedlar	DuPont Tedlar	Polyvinyl fluoride film, non-oriented	Surfacing, weathering film
Teflon	DuPont Films	Polytetrafluoroethylene	High temperature material
FEP	DuPont Films	Tetrafluoroethylene-hexafluoropropylene copolymer	
Teflon PFA	DuPont Films	Perfluoroalkoxy-fluorocarbon copolymer resin	High performance film
Tefzel	DuPont Films	Ethylene-tetrafluoroethylene copolymer resin	High performance film

Trade name	Manufacturer	Chemical identification	Heat deflection temperature (264 psi) or application
Tenite	Eastman Chemical	Cellulose acetate resin	111–196°F
	Eastman Chemical	Cellulose acetate butyrate resin	113–201°F
	Eastman Chemical	Cellulose propionate resin	108–228°F
Tenite	Eastman Chemical	(Linear) low density polyethylene resin	(Food) packaging
Tenite PET	Eastman Chemical	(Crystallizable) polyethylene terephthalate resin	(Food) packaging
Terblend	BASF	Acrylic ester-styrene-acrylonitrile terpolymer/PC alloy	
Terluran	BASF	Acrylonitrile-butadiene-styrene terpolymer	
Terlux	BASF	Clear acrylonitrile-butadiene-styrene resin	
Texin	Bayer Corp.	Thermoplastic polyurethane extrusion resin	
Thermx	Eastman Chemical	Copolyester resin, superior heat and impact	Ovenable food trays
Thermx PCT	Eastman Chemical	Glass fiber filled cyclohexylenedimethylene terephthalate	480°F, structural sheet resin
Topas	Ticona Corp.	Cyclic olefin copolymers	167–338°F (66 psi)
Torayfan	Toray Plastics	Biaxially oriented, treated polypropylene film and sheet	
Torlon	Amoco Performance Prod.	Polyamide-imide	535°F, tooling
TPX	Mitsui Petrochemical	Methylpentene copolymer	Food packaging
Traytuf	Shell Chemical	Polyethylene terephthalate extrusion resin	
Triax	Bayer Corp.	Acrylonitrile-butadiene-styrene blend	
Trycite	Dow Plastics	Biaxially oriented polystyrene film	Packaging
Tuffak	AtoHaas	Polycarbonate sheet	
Twintex	Vetrotex CertainTeed	Drapable glass-fiber rowing and thermoplas. filament sht.	
Tyril	Dow Plastics	Styrene-acrylonitrile copolymer resin	
Udel	Amoco Performance Prod.	Polysulfone resin	217°F
Ultem	GE Plastics	Polyetherimide resin and sheet	345°F
Ultradur	BASF	Polybutylene terephthalate resin	375–408°F
Ultralite	O'Sullivan	Press polished polyvinyl chloride sheet	160°F
Ultramid	BASF	Polyamide 6, 66, and 6/10	

(continued)

Trade name	Manufacturer	Chemical identification	Heat deflection temperature (264 psi) or application
Ultrapek	BASF	Polyaryletherketone resin	
Ultrason	BASF	Poly (ether)sulfone resins	
Ultrathene	Millennium Petrochemicals	Polyethylene and ethylene-vinyl acetate copolymer	Packaging and heat sealing
Unichem	Colorite Plastics	Polyvinyl chloride compounds	
Unite	Aristech Chem. Corp.	Maleic anhydride grafted polypropylene	Chemical coupling layer
Unoflex	Morton International		Laminating adhesive
Valox	GE Plastics	Polybutylene terephthalate resin	130–210°F
Valtek	Montell North America	Polypropylene	
Vespel	DuPont Engg. Polymers	Polyimide molded material	High temperature material, tooling
Victrex	Victrex USA Inc.	Polyether-etherketone, polyethersulfone resin	320°F, 397°F
Vinika	Schulman	Polyvinyl chloride compounds	
VistaFlex	Advanced Elastomer	Thermoplastic elastomer, plasticized PVC replacement	
Vitafilm	Huntsman Packaging	Extruded polyvinyl chloride film	
Vivak	Sheffield Plastics	Extruded copolyester PETG sheet	
Vydyne	Solutia	Polyamide 66	
Vynathene	Millennium Petrochemicals	High vinyl acetate-ethylene copolymer	Adhesive resin
Winwrap	Huntsman Packaging	Cast coextruded polyethylene stretch film	Packaging
Xenoy	GE Plastics	Polycarbonate/polybutylene terephthalate alloy	205°F
XT Polymer	Cyro Industries	Acrylic-based multipolymer compound	186–194°F
X-TC	DuPont Automotive	Long glass-fiber-reinforced polyethylene terephthalate	494°F, structural sheet
Xydar	Amoco Performance Prod.	Aromatic copolymer (liquid-crystal-polymer)	671°F
Zytel	DuPont Engg. Polymers	Polyamide 66 resin	158–212°F, 425–474 (66 psi)
Zytel ST	DuPont Engg. Polymers	Polyamide/ethylene-propylene-diene alloy	

*Many of the listed proprietary names are registered trademarks in the United States.

Listing of materials manufacturers.

Name	Address	Telephone
Advanced Elastomer Systems	388 S. Main St., Akron, OH 44311-1059	(800) 352-7866
Albany International Corp.	777 West St., Mansfield, MA 02048-9114	(508) 339-7300
AlliedSignal Inc.	101 Columbia Rd., Morristown, NJ 07962	(800) 821-9292
Alucobond Technologies, Inc.	P.O. Box 507, Benton, KY 42025	(800) 626-3365
American Mirrex Corp.	1389 School House Rd., New Castle, DE 19720	(800) 488-7608
Amoco Chemicals	200 E. Randolph Dr., Chicago, IL 60601-7125	(800) 621-4590
Amoco Performance Products	4500 McGinnis Ferry Rd., Alpharetta, GA 30202-3914	(800) 621-4557
ARCO Chemical Co.	3801 West Chester Pike, Newtown Square, PA 19073-2387	(610) 359-5642
Aristech Chem. Corp.	7350 Empire Dr., Florence, KY 41042	(800) 354-9858
AtoHaas North America Inc.	Independence Mall West, Philadelphia, PA 19105	(215) 785-8290
Ausimont USA, Inc.	44 Whippany Rd., Morristown, NJ 07962	(800) 221-0553
Azdel, Inc.	925 Washburn Switch Road, Shelby, NC 28150	(810) 351-8000
Bamberger Polymers, Inc.	1983 Marcus Ave., Lake Success, NY 11042	(800) 888-8959
BASF Plastics Materials	3000 Continential Dr. North, Mount Olive, NJ 07828-9909	(800) 227-3746
Bayer Corp.	100 Bayer Rd., Pittsburgh, PA 15205-9741	(800) 622-6004
BFGoodrich Specialty Chemicals	9911 Brecksville Rd., Cleveland, OH 44141-3247	(800) 331-1144
BOC Coating Technology	4020 Pike Lane, Concord, CA 94520-1297	(510) 680-0501
Borden Packaging Div. Borden Inc.	One Clark St., North Andover, MA 01845	(508) 686-9591
BP Chemicals Inc., Barex Div.	4440 Warrensville Ctr. Rd., Cleveland, OH 44128	(800) 272-4367
Charlotte Chemical Inc.	250 Wilcrest, Suite 300, Houston, TX 77042	(713) 954-4855
Chisso America, Inc.	1185 Ave. of the Americas, New York, NY 10036	(212) 302-0500
Colorite Plastics Co.	101 Railroad Ave., Ridgefield, NJ 07657	(201) 941-2900
CYRO Industries	P.O. Box 5055, Rockaway, NJ 07866	(800) 631-5384
Dow Chemical Company	2040 Dow Center, Midland, MI 48674	(800) 441-4369
DSM Engineering Plastics	2267 West Mill Rd., Evansville, IN 47732-3333	(800) 438-7225
DuPont Automotive	950 Stephenson Highway, Troy, MI 48007-7013	(810) 583-8000
DuPont Corian Products	Barley Mill Plaza, Wilmington, DE 19880	(302) 992-2072
DuPont Engineering Polymers	Chestnut Run, Wilmington, DE 19805	(800) 441-0575

(continued)

221

Name	Address	Telephone
DuPont Films	Barley Mill Plaza, Wilmington, DE 19805	(810) 237-4357
DuPont Packaging and Industrial Polymers	P.O. Box 80010, Wilmington, DE 19898	(800) 441-7515
DuPont Tedlar	Sheridan Drive and River Rd., Buffalo, NY 14207	(800) 255-8386
Eastman Chemical Co.	P.O. Box 431, Kingsport, TN 37662-5280	(800) 327-8626
Elf Atochem North America, Inc.	2000 Market St., Philadelphia, PA 19103	(800) 225-7788
Emabond Systems, Ashland Chem. Co.	49 Walnut St., Norwood, NJ 07648	(201) 767-7400
Emerson & Cuming	59 Walpole St., Canton, MA 02021	(617) 821-4250
EMS-American Grilon	2060 Corporate Way, Sumter, SC 29151-1717	(803) 481-9173
EVAL Company of America	1001 Warrenville Rd., Lisle, IL 60532-1359	(800) 423-9762
Exxon Chemical Co.	13501 Katy Freeway, Houston, TX 77079-1398	(800) 231-6633
Ferro Corp., Filled & Reinforced Plastics Div.	5001 O'Hara Dr., Evansville, IN 47711	(812) 423-5218
Flex-O-Glass, Inc.	1100 N. Cicero Ave., Chicago, IL 60651	(312) 379-7878
Garland Mfg. Co.	P.O. Box 538, Saco, ME 04072-0538	(207) 283-3693
GE Plastics	One Plastics Ave., Pittsfield, MA 01201	(800) 437-5278
Geon Co., The	6100 Oak Tree Blvd., Cleveland, OH 44131	(800) 438-4366
Goodyear Tire & Rubber Co., Chemical Div.	1485 E. Archwood Ave., Akron, OH 44316	(800) 522-7659
Hampshire Chemical Corp.	55 Hayden Ave., Lexington, MA 02173	(617) 861-9700
Huntsman Packaging Corp.	3575 Forest Lake Dr., Uniontown, OH 44685	(800) 321-2385
ICI Acrylics Inc.	10091 Manchester Rd., St. Louis, MO 63122	(800) 325-9577
Kama Corp.	666 Dietrich Ave., Hazleton, PA 18201	(717) 455-2021
Kleerdex Co.	P.O. Box 3248, Aiken, SC 29802	(800) 325-3133
Klockner Pentaplast of America, Inc.	Klockner Rd., Gordonsville, VA 22942	(540) 832-3600
M.A. Industries Inc., Polymer Div.	303 Dividend Dr., Peachtree City, GA 30269	(800) 241-8250
Millennium Petrochemicals	P.O. Box 429550, Cincinnati, OH 45249	(800) 323-4905
Mitsubishi Gas Chemical America, Inc.	520 Madison Ave., New York, NY 10022	(212) 752-4620
Mitsui Plastics, Inc.	11 Martine Ave., White Plains, NY 10606	(914) 287-6800

Name	Address	Telephone
Mobil Chemical Co.	P.O. Box 3029, Edison, NJ 08818-3029	(908) 321-6000
Montell North America	2801 Centerville Rd., Wilmington, DE 19850-5439	(302) 996-6000
Morton International, Inc.	100 N. Riverside Plaza, Chicago, IL 60606-1598	(312) 807-3106
Nova Chemicals Inc.	400 Frankfort Rd., Beaver Valley Site; Monaca, PA 15061-2298	(412) 773-5633
O'Sullivan Corp.	1944 Valley Ave., Winchester, VA 22601	(800) 336-9882
Phillips Chemical Co.	P.O. Box 58966, Houston, TX 77258-8966	(800) 231-1212
Plastic Suppliers	2887 Johnstown Rd., Columbus, OH 43219	(800) 722-5577
Polycast Technology Corp.	70 Carlisle Place, Stamford, CT 06902	(800) 243-9002
Polymer Composites, Inc.	P.O. Box 30010, Winona, MN 55987-1010	(507) 454-4151
Polymer Corp.	2120 Fairmont Ave., Reading, PA 19612	(800) 729-0101
Polymer Extruded Products, Inc.	297 Ferry St., Newark, NJ 07105	(201) 344-2700
Quadrax Corp.	300 High Point Ave., Portsmouth, RI 02871	(401) 683-6600
Rexene Products Co.	5005 LBJ Freeway, Dallas, TX 75244	(800) 695-2985
Reynolds Metals Co.	6601 W. Broad St., Richmond, VA 23230	(804) 281-2000
Schulman, A., Inc.	3550 W. Market St., Akron, OH 43693	(800) 547-3746
Segen, Edward, & Co., Inc.	11 Kent St., Milford, CT 06460	(203) 878-6503
Sheffield Plastics, Inc.	119 Salisbury Rd., Sheffield, MA 01257	(800) 628-5084
Shell Chemical Co.	P.O. Box 2463, Houston, TX 77252-2463	(800) 457-2866
Solutia	800 N. Lindbergh Blvd., St. Louis, MO 63167	(573) 694-1000
Solvay Polymers, Inc.	P.O. Box 27328, Houston, TX 77227-7328	(800) 231-6313
Ticona Corp.	90 Morris Ave., Summit, NJ 07901	(800) 526-4960
Toray Plastics (America), Inc.	50 Belver Ave., North Kingstown, RI 02852-7500	(401) 294-4511
Toyo Seikan Kaisha, Ltd.	444 N. Michigan Ave., Chicago, IL 60611	(312) 822-0444
Tredegar Film Products	1100-T Boulders Pkwy., Richmond, VA 23225-4035	(804) 330-1222
Twinpak Inc.	1840 Trans-Canada Hwy., Dorval, Quebec, Canada H9P 1H7	(514) 684-7070
Uniroyal Technology Corp., Royalite Div.	312 N. Hill St., Mishawaka, IN 46546-0568	(219) 259-1259
Vetrotex	P.O. Box 860, Valley Forge, PA 19482	(800) 274-8530
Victrex USA Inc.	601 Willowbrook Lane, West Chester, PA 19382	(610) 696-3144
Westlake Plastics Co.	P.O. Box 127, W. Lenni Rd., Lenni, PA 19052	(215) 459-1000

Appendix D

Conversion factors

THE FOLLOWING PAGES present conversion factors for property values listed in Tables 2.1, 2.2, 9.1, Appendix A, and Appendix B. At the end of this appendix is a temperature conversion table.

Table 2.1 (page 16)

Specific gravity: $1 \dfrac{g}{cm^3} = 62.4 \dfrac{lb}{cu\,ft}$

Specific heat: $1 \dfrac{Btu}{lb\,°F} = 1 \dfrac{cal}{g\,°C}$

Heat of fusion: $1 \dfrac{Btu}{lb} = 1 \dfrac{cal}{g}$

Thermal conductivity: $1 \dfrac{Btu\,ft}{sq\,ft\,hr\,°F} = 12 \dfrac{Btu\,in.}{sq\,ft\,hr\,°F} = 0.00413 \dfrac{cal\,cm}{cm^2\,sec\,°C} = 0.0173 \dfrac{W\,cm}{cm^2\,°C}$

Linear coefficient of thermal expansion: $1 \dfrac{in.}{in.\,°F} = 1.80 \dfrac{cm}{cm\,°C}$

Table 2.2 (page 28)

Heat deflection temperature: 264 psi = 1.82 MPa = 1.82 N/mm²

66 psi = 0.455 MPa = 0.455 N/mm²

Temperature conversion: °F = $\frac{9}{5}$ °C + 32

Temperature conversion table: see page 229

Table 9.1 (page 148)

Gas transmission rate (gas permeability):

$$1 \frac{cm^3 \, mil}{100 \, in^2 \, 24 \, hr} = 15.50 \frac{cm^3 \, 25 \, \mu m}{m^2 \, 24 \, hr \, atm} = 151.7 \frac{cm^3 \, 25 \, \mu m}{m^2 \, 24 \, hr \, MPa} = 2.006 \cdot 10^{-18} \frac{mol \, (STP)}{m \, sec \, Pa}$$

Moisture vapor transmission rate (moisture vapor permeability):

$$1 \frac{g \, mil}{100 \, in^2 \, 24 \, hr} = 15.50 \frac{g \, 25 \, \mu m}{m^2 \, 24 \, hr} + 0.393 \frac{g \, mm}{m^2 \, 24 \, hr} = 0.926 \frac{grains \, mil}{sq \, ft \, hr}$$

Table Appendix A (page 202)

Flexural modulus of elasticity: $1000 \dfrac{lb}{sq\ in.} = 6.895\ MPa$

Impact strength: $1 \dfrac{ft\ lb}{in.} = 53.35\ \dfrac{J}{m}$

Temperature: see above
Specific heat: see above
Thermal conductivity: see above
Linear coefficient of thermal expansion: see above
Specific gravity: see above

Price: $1\ \dfrac{\$}{lb} = 2.20\ \dfrac{\$}{kg}\ ;\ 1\ \dfrac{\$}{sq\ ft} = 10.76\ \dfrac{\$}{m^2}$

Relative Price: $1\ \dfrac{¢}{cu\ in.} = 0.0610\ \dfrac{¢}{cm^3}$

Table Appendix B (page 206)

Tensile strength: $1000 \frac{lb}{sq\ in.} = 6.895\ MPa$

Area factor: $1 \frac{sq\ ft\ mil}{lb} = 144 \frac{sq\ in.\ mil}{lb} = 0.202 \frac{m^2\ 25\ \mu m}{kg} = 0.00505 \frac{m^2\ mm}{kg}$

Burst strength: 1 Mullen point = 1 psi = 6.895 KPa

Gas transmission rate: see above

Moisture vapor transmission rate: see above

Price of film: $1 \frac{\$}{lb} = 2.20 \frac{\$}{kg}$

Temperature conversion table.

°F	°C	°F	°C	°F	°C	°F	°C
50	10	200	93.3	350	176.7	500	260
55	12.8	205	96.1	355	179.4	510	265.6
60	15.6	210	98.9	360	182.2	520	271.1
65	18.3	215	101.7	365	185	530	276.7
70	21.1	220	104.4	370	187.8	540	282.2
75	23.9	225	107.2	375	190.6	550	287.8
80	26.7	230	110	380	193.3	560	293.3
85	29.4	235	112.8	385	196.1	570	298.9
90	32.2	240	115.6	390	198.9	580	304.4
95	35	245	118.3	395	201.7	590	310.0
100	37.8	250	121.1	400	204.4	600	315.6
105	40.6	255	123.9	405	207.2	610	321.1
110	43.3	260	126.7	410	210	620	326.7
115	46.1	265	129.4	415	212.8	630	332.2
120	48.9	270	132.2	420	215.6	640	337.8
125	51.7	275	135	425	218.3	650	343.3
130	54.4	280	137.8	430	221.1	660	348.9
135	57.2	285	140.6	435	223.9	670	354.4
140	60	290	143.3	440	226.7	680	360.0
145	62.8	295	146.1	445	229.4	690	365.6
150	65.6	300	148.9	450	232.2	700	371.1
155	68.3	305	151.7	455	235	710	376.7
160	71.1	310	154.4	460	237.8	720	382.2
165	73.9	315	157.2	465	240.6	730	387.8
170	76.7	320	160	470	243.3	740	393.3
175	79.4	325	162.8	475	246.1	750	398.8
180	82.2	330	165.6	480	248.9	760	404.4
185	85	335	168.3	485	251.7	770	410.0
190	87.8	340	171.1	490	254.4	780	415.6
195	90.6	345	173.9	495	257.2	790	421.1
200	93.3	350	176.7	500	260	800	426.7

For each °F increment add °C:

1	0.6
2	1.1
3	1.7
4	2.2

Index*

ABS, *see* acrylonitrile-butadiene-styrene copolymer
Absorption, 5, 12
AC Technology North America, Inc., 46
Acclear, C210
Accpro, C210
Acctuf, C210
Accuform, 47
Acetal, 141, 156, C212
Aclar, C210
Acrylics, T8, T16, 49, 115, 130, 136, 145, C210, C212, C214, C216–218, C220
Acrylite, C210
Acrylonitrile-butadiene-stryrene copolymer, T28, 31, 128, 130, 140, 144, 145, 195, T202, C211, C214–217, C219
Acrylonitrile-styrene-acrylate copolymer, 130, C211, C213, C214
Acrysteel, C210
Acrystone, C210
Adcote, C210
Adheflon, C210
Admer, C210
Advanced Elastomer Systems, C217, C220, 221
Affinity, C210
Ageless, 149, C210
Aging of plastics, 131
Aheflon, T151
AIM, C210
Air convection losses, 25
Air drafts, 26, 27, 51
Air Products and Chemicals, Inc., 149
Air slip forming, 167
Airopak, 149
Akuloy, C210
Albany International Corp., C217, 221

Allen-Bradley Co., 32
AlliedSignal Inc., 141, C210, C211, C214, C216, 221
Alloys, **143**, 144
Altair, C210
Alucobond Technologies, Inc., C213, C218, 221
Aluminum, T16
American Mirrex Corp., C218, 221
Amoco Chemicals, 149, C210, 221
Amoco Performance Products, C210, C214, C215, C217, C219, C220, 221
Amodel, C210
Amorphous, polymers, **111**, 113, 120, 122, 126, T151, 217
  ethylene terephthalate copolyester, T29, 124, 126, C217
Amosorb, 149, C210
Annealing, 77, 173, 187, 190
Apec, C210
Appeel, C210
Arcel, C210
ARCO Chemical Co., C212, 221
Ardel, C210
Area factor of films, T206
Areal draw ratio, 36
Aristech Chemical Corp., T152, 189, C210, C214, C217, C220, 221
Ashland Chemical Co., 186
Atlas Vac Machine, 81, 88
AtoHaas North America Inc., C213, C214, C216, C219, 221
Attane, C210
Ausimont USA, Inc., C213, 221
Azdel, Inc., C210, C211, 221
Azloy, C210
Azmet, C211

*Page numbers for the main treatment of a subject are in **bold face** when more than one reference is given.

Index

Bamberger Polymers, Inc., C211, 221
Bapolan, C211
Bapolene, C211
Bapolon, C211
Barex, T151, C211
Barrier materials 1, 60, 105, 124, 130, 139, **146–152**, 194, 198, C210–218
BASF Plastics Materials C214, C215, C218, C219, C220, 221
Bayblend, C211
Bayer Corp., C210, C211, C212, C214, C215, C219, 221
Bexloy W, C211
BFGoodrich Specialty Chemicals, C213, C218, 221
Biaxial orientation, 46, 115, **123**, 129, 193, 197
Biaxially oriented films, T28, 140, T148, 168, T206, C214
Biaxially oriented sheet, 37
Biaxially stretched, 36
Bicor, C211
Billow forming, 37, 60, 64, 159, 165
Biodegradable plastics, 155, 157, C211
Biomax, C211
Blends, *see* alloys and polymers
Blister formation, 27, 120
Blister packaging equipment, 104
Blow molding processes, 46, 171, 199
BOC Coating Technology, 149, C217, 221
Bonding, 185
Borden Packaging, C217, 221
BP Chemicals Inc., Barex Div., T151, C211, 221
Brittle-ductile fracture transition, 193
Brown Machine, 82, 85
Bubble thermoforming, 159
Bulk density, T16
Bullet-proof laminates, 121
Burst strength of films, T206
Bynel, T151, C211

C-MOLD Thermoforming, 46
CA, *see* cellulose acetate
CAB, *see* cellulose acetate-butyrate
Cadon, C211
Calendering, 114, 128
Calibre, C211
Calrod heaters, 25
CAP, *see* cellulose acetate-propionate
Capran, C211
Capron, C211
Carbon dioxide, 146, T148

Carboxyl groups, chemical component of polyesters, 142
Cargill Inc., 155
Carilon, C211
Cascading, *see* recycling
Casting process, 129
Catalytic gas heaters, 23
Cavity forming, *see also* female mold, 35, 161
Cell cast acrylic sheets, 136
Cellophane, 149, T206
Celluloid, 2
Cellulose acetate, T8, T28, **136**, 145, T148, T202, T206, C219
Cellulose acetate-butyrate, T28, **136**, 145, 196, T202, T206
Cellulose (acetate-)propionate, T28, 54, **136**, 145, T202, C219
Cellulosics, T28, 21, 37, 49, 76, 114, 117, 119, 120, **136**, **137**, 161
CelsiStrip., 30
Celstran, C211
Centrex, C211
Charlotte Chemical Inc., C214, 221
Chemical vapor deposition, C217
Chill roll casting, 128
Chill-marks, 53, 70, T181
Chiller, 71
Chisso America, Inc., C212, 221
Clamping frame, 33
Cleartuf, C211
Clysar, C211
Co-Pak process, 197
Coefficient of thermal expansion, linear, *see* thermal expansion
Coextrusion, 129, 130, **149–152**, 198, C215
COFO process, 195
Cold forming, 122, 195
Cold stamping, 195
Colorite Plastics Co., C220, 221
Competing forming processes, 193
Compression molding, 53
Computer aided engineering, 45
Computer-integrated manufacturing, 32
Conduction of heat, 3, 15, 20, 69
Continental Can Co., 194
Continuous in-line thermoformer, 85
Continuous rotary thermoform, 100
Controls, 32, 103, **109–110**
Convection heating, 3, 4, **18**, 26
Convection cooling, 69
Cooling of part, **69–71**
Cooling fixtures, 50

Index

Copolymers, **143**
Copper, T16
Corian, 124, C211
Corterra, C211
Cost of materials, **131–133**, 184, T202
Cotie, T151, C211
CPET, *see* (crystallisable) polyethylene terephthalate
Crazing, 67, 122, T180
Creep failure, 131
Cross-linking, 43, 72, 115
Crystalline materials, 21, 49, 50, 53, 71, 115, 117, 124, 137, C211
Crystalline melting point, 112, 138, 194, T202
Crystallizable polyethylene terephthalate, T29, 72, 124, C211
Crystallization, 72, **120–127**
Cuspation-dilation forming process, 194
Cyclic olefin copolymers, C219
Cyclohexylenedimethylene phthalate polyesters, 124, 142, C212
Cycolac, C211
Cycoloy, C211
CyRex, C212
Cyro Industries, C210, C212, C220, 221
Cyrolite, C212
Cyrolon, C212

Dancer rolls, 97
Daran, C212
Daratak, C212
Deburring, 76, 77
Decorating thermoformed parts, **190**
Delrin, C212
Denesting lugs, 51, 191
Density, *see also* specific gravity, T16, 118, 125, 138, T203
Depth of draw, 37
Design considerations, 183
Diaphragm forming, 195
Dielectric heating, 4, 67
Diffusivity, *see* permeability
Dimensional tolerances, 30, 35, **48**, 66, 117
DKE, C212
Dow Chemical Company, 194, C210–212, C214–217, 221
Dow Plastics, 122, T151, C216, C218, C219
Dowlex, C212
Draft in the mold, **52**, T180, 185
Drape forming, *see also* male molds, 35, 67, 80, 144, 162
Draw ratio, **35**, **36**, 43

Draw down during extrusion, 49, 117
Drawing, mechanical forming, 68
Drying of sheet, *see also* water absorption, 4, 119
DSM Engineering Plastics, C210, C218, 221
Ductile-brittle fracture transition, 193
DuPont Automotive, C211, C214, C220, 221
DuPont Corian Products, 124, C211, 221
DuPont Engineering Polymers, C212, C217, C220, 221
DuPont Films, T151, C211, C214, C215, C218, 222
DuPont Packaging, T151, C210, C211, C213, C215, C217, C218, 222
DuPont Tedlar, C218, 222
Durethan, C212
Dylark, C212
Dylene, C212
Dylite, C212

Eastalloy, C212
Eastapak, C212
Eastar, C212
Eastman Chemical Co., 125, C212, C215, C218, C219, 222
Eccentric linkages, 68
EcoPLA, 155
Edge canting, 173
Edge losses of heat, 13
Elastic deformation, 46, 113
Elastic springback, 196
Elastomer modified styrene maleicanhydride resin, C218
Electrical properties, 152
Electromagnetic interferences, 153
Elf Atochem North America, T151, C210, C214, C216, C217, 222
Elite, C212
Elongation at break of films, T206
Emabond Systems, Ashland Chem. Co., 186, C212, 222
Emaweld, C212
Emerson & Cuming, C218, 222
EMI shielding plastics, 153
Emissivity, 5, 7, 10, 27, 30
EMS-America Grilon, C213, 222
End-use temperature, testing at, 49
Enduran, 124, C212
Energy density of heaters, 21–23
Engineering plastics, 23, 71, 122, **141**, T202
Environmental stress cracking, 77, 131
Epolene, C212

Index

Epoxy resins, T16, 54, 55
Escorene, C212
Ester groups, chemical component of polyesters, 142
Ethafoam, C212
Ethylene-chlorotrifluoroethylene copolymer, 141, T206, C213
Ethylene glycol, chemical component of polyesters, 123
Ethylene-vinylacetate copolymer, **139**, 145, 153, T202, T206, C210–212, C220
Ethylene-vinylalcohol copolymer, 60, 139, T148, T151, 194, C214, C218
Ethylene-methacrylic acid copolymer resin, *see also* ionomer, C215, C218
Ethylene-tetrafluoroethylene copolymer resin, C218
EVAL Company of America, T151, C212, C222
Everflex, C212
Exceed, T151
Expan, C212
Expanded polystyrene sheet, 175, C212, C216
Extruder specification, 98
Extrusion compounds, superiority of, 114, 128, 198
Exxon Chemical Co., C212, 222

Fatigue failure, 131
Female funnel mold, 37
Female molds, 35, **39**, 43, 48, 49, 52, 71, 104, **161**
FEP, *see* fluorinated ethylene-propylene copolymer
Ferro Corp., C213, C216, 222
Fiber reinforced prepreg sheets, 144, 176, C217
Fiber reinforced structural supports, 188
Finishing formed parts, 153, 190
Fire risk, 23
Flair, C213
Flame retardant compounds, 140, 156, C213
Flammability of plastics, 158
Flamtard, C213
Flex-O-Glass, Inc., C218, 222
Flow-forming process, 176
Fluid pressure forming, 175
Fluid-jet trimming, 85
Fluorinated ethylene-propylene copolymer, T29, 141, 145, T148, T202, T206, C218
Fluorocarbon polymers, 3, 7, T9, 49, 141, C210, C211, C213, C218
Fluorosint, C213

Foamed products
  polycarbonate, C213
  polyethylene, C210, C212, C215
  polystyrene, 4, 68, 140, C212, C213, C218
  polyurethane, C214
  polyvinyl chloride, C218
Foamed sheet forming, 97
Foaming-in-place process, 188
Fome-Cor, C213
Food packaging, 103, 105, 124, 130, 146
Forex EPC, C213
Formion, C213
Fortex, T151
Fortiflex, C213
Fortilene, C213
Fortron, C213
Four station rotary thermoformer, 169, 170
Free forming, 18, 37, 70, 159
Frozen-in stresses, 34, 128
Fusabond, C213

Gabler Maschinenbau, GmbH, 75, 96
Gapex, C213
Gar-Dur, C213
Garland Mfg. Co., C213, 222
Gas transmission rate, T148, T206
Gas-fired infrared heaters, 22
GE Plastics, 124, T180, C211–215, C218–220, 222
Geloy, C213
Geon Co., The, C213, 222
Glass fiber-reinforced structures, **144**, 189
  amorphous engineering plastics, C210
  semicrystalline engineering plastics, C211
  thermoplastics, 144, C210, C211, C213, C219, C220
  unsaturated polyester, 189
Glass transition temperature, 111, 113, 120, 121, 127, 142
Gloss, 53, 129
Goodyear Tire & Rubber Co., C216, 222
Graphite, T16
Gravitational force, 66, 112
Grid-assist forming, 43
Grilamid, C213
Grilbond, C213
Grilon, C213
Grivory, C213

Halar, C213
Halogen containing plastics, T16, T29, 140, 141, T202, T206

Index **235**

Halogen lamps, 24
Hampshire Chemical Corp., C212, 222
Hardwood, T16, 54, 60, 70
HDPE, *see* high-density polyethylene
Heat, *see also* thermal
Heat deflection temperature, T28, **112**, T202
Heat losses, 12, 13
Heat of fusion, T16, 18, 70, **117**
Heat penetration, 10, 24, 116
Heat profiling, 25
Heat sealing, **104–108**, 130, 149, 198, C210, C216
Heat sink compound, 70
Heat transfer liquid, 4
Heat-shrink film, 127, C215
Heater
    arrangement, 18, 26
    control, 92
    efficiency, 10, 23, 25
Heating of film and sheet, **3–15**
Heating time, 21, 24, 25
Heating tunnel, 26, 85, 92
Heavy gauge sheet
    heating, 4, 18, 22, 24, 27, 83
    drying, 119
    cooling, 69
Heavy-gauge sheet, definition, 2
HiFax, C213
High-density polyethylenes, 137, T202
High-impact polystyrene, T28, **140**, 145, T202
High-temperature plastics containing ring structures, 141
HiGlass, C213
HIPS, *see* high-impact polystyrene
Hivalloy W, C213
Hot strength of plastics, 112
Huntsman Packaging Corp. C220, 222
Hycar, C213
Hydraulically operated thermoformer, 68, 103
Hydrogen bonds, 141
Hydrolysis, chemical breakdown, 115, 120
Hydroxyl group, chemical component of polyesters, 142
Hyflon MFA, C213
Hygard, C213
Hylar, C213
Hyrizon, C214
Hyzod, C214

ICI Acrylics Inc., C214, C216, 222
Illig Maschinenbau GmbH, 95–97, 189
Impact strength, 130, 134, 138, 142, 144, T202

Impax, C214
Implex, C214
Inert gas atmosphere for packaging, 104, 106
Infrared absorption, 3, T8, T9
Infrared heaters, 10, **20–24**, 79
Infrared sensors, 27, 30, 32
Injection molding, 50, 53, 66, 114, 118, 172, 198
Inline thermoformer, 95, 98
Inserts, 186
Insufficient draw, T180
Ionomer resin, 139, 145, T206, C211, C213
Irwin Research & Development, Inc., 89–91, 94
Irwin International, 100, 102
Izod impact strength, T202

K-resin, C214
Kadel, C214
Kaladar, C214
KamaCorp., C214, 222
Kapton, C214
Keldax, C214
Kiefel, GmbH, 15, 74, 75, 100, 105, 106
Kleerdex, Co., C214, 222
Klockner Pentaplast, C216, 222
Korad, 130, C214
Kraemer & Grebe GmbH, 107, 108, 109
Kraton, C214
Krystaltite, C214
Kydex, C214
Kynar, C214

Lamal, T151, C214
Lamicon, T151, C214
Laminar blend barrier plastic, T151, C217
Laminations, 129, 149
Lamisan, C214
Landfills, 155
Laser beam cutting, 74, 77
LDPE, *see* low-density polyethylene
Lethery state of plastics, 111
Lennite, C214
Lexan, C214
Linear bending, 115
Linear Form Pty. Ltd. 100, 101
Linear low-density polyethylene, 137
Linear thermoformers, 100
Lip rolling, 76, 93
Liquid carbon dioxide for cooling, 72
Liquid-crystal polymer, C220
LLDPE, *see* linear low-density polyethylene
Low-density polyethylene, 137, T202
Lucite, C214

Lumirror, C214
Luran, C214
Lustran, C214

M.A. Industries Inc., C215, 222
MAAC Machinery Corporation, 171
Machining of parts, 76
Magnum, C214
Mahaffy & Harder Engineering Co., 106
Makroblend, C215
Makrofol, C215
Makrolon, C215
Male molds, 27, 35, **37**, 41, 48, 49, 52, 64, 69, 117, **162**
Maleic anhydride modified polyolefin, C213, 220
Malon, C215
Marbach Tool & Equipment, Inc., 52
Mark-off part defects, 22, 39, 60
Marlex, C215
Matched-mold forming, 80, 144, 175
Material properties, 111–133
Mechanical forming, 61, 66, 172
Mechanically operated thermoformers, 103
Melinar, C215
Melinex, C215
Melt flow stamping, 176
Melt index, 137
Melt phase forming, 60, 139
Melting temperature, T28, **111**, 120, 122, 124, 144, T202
Metapor, C215
Micro-Phaser, 87, 93
Microprocessors, 110
Microwave heating, 4, 67
Millennium Petrochemicals, T151, C216, C220, 222
Mindel, C215
Mineral filled polybutylene terephthalate, 124, C212
Mitsubishi Gas Chemical Co., 149, T151, C210, C215, 222
Mitsui Petrochemicals, C210, C219
Mitsui Plastics, Inc., C214, 222
Mobil Chemical Co., C211, C216, 222
Modic, T151, C215
Modulus of elasticity in flexure (stiffness), 113, **130**, 194, T202
Moisture, *see also* water
Moisture effect on electrostatic discharge, 153
Moisture vapor transmission rate, T148, T206

Moisture-scavenging layer in packaging films, 149
Mold cooling, 56
Mold construction, **35–59**
Mold materials, 54
  hardwood, T16, 54, 60, 70
  metal, 55
  plastic, 70
Mold plug material, C218
Mold plugs, 59
Mold temperature, T28, 48, 70, 71
Molded-in stresses, 53
Molding shrinkage, 49, 71
Monaformer, 196
Monax, C215
Monsanto, Co., C213, C220, 223
Montell North America, C213, C215, C216, C220, 223
Moplen, C215
Morton International, Inc., T151, T152, C210, C214, C218, C220, 223
MXD6, T151, C215
Mxsten, C215
Mylar, T151, C215

National Can Co., 197
Necking of sheet, T181
Neopolen, C215
Non-melt forming process, 194
Noryl, C215
Nova Chemicals Inc., C210, C212, 223
Novodur, C215
Novolen, C215
Nucrel, C215
Nylon, T9, T29, 112, 115, 119, 120, 122, 141, 142, 145, T148, C210–213, C215, C217–220
Nylopak, C215

O'Sullivan Corp. C217–219, 223
Okura Industrial Co., T151
Opaque plastics, 117, 122, 125, 138, 145
Oppalyte, C216
OPS, *see* oriented polystyrene
Optical properties, 57, 129, 139, **145**, 159
Optum, C216
Orientation, 36, 49, 60, 115, **120–123**, 136, 141, 147
Oriented films, T206
  high-density polyethylene, C215
  polypropylene film, C211, C216
  polystyrene film, T28, 49, 76, C216, C219

Index

Oxygen permeability, 130, T148
Oxygen absorber, scavenger, 148, 149, C210, C216
Oxyguard, 149, C216
Oxyshield, C216

PA, *see* nylon (polyamide)
Packaging container forming, 196
PAI, *see* polyamide-imide
PC, *see* polycarbonate
PC Materials, Inc., 149
PCT, PCTA and PCTG, *see* ethylene terephthalate (co)polyesters
Pebax, C216
Pelaspan-Pac, C216
Pellethane, C216
PEN, *see* polyethylene naphthalate
Pentafood, C216
Pentaform, C216
Pentamed, C216
Pentapharm, C216
Permeability, 146
Perspex, C216
PES, *see* polyether sulfone
Petainer Development Company, 194
PET, *see* polyethylene terephthalate
PETG, *see* amorphous ethylene terephthalate copolyester
Petlite II, C216
Petrothene, C216
Phenolics, T16, 54, 115
Phillips Chemical Co., C214, C215, C217, 223
Photoelectric cell, 30
Pigmentation, 10
Plasma chemical vapor deposition, 149
Plastic memory, 50, 121
Plastic suppliers, C216, 223
Plastisol molding, 199
Plexar, T151, C216
Plexiglas, C216
Pliovic, C216
Plug-assists forming, T28, 37, 59, 80, 87, 93, 164
Plug-and-ring forming, 173
PMMA, *see* polymethyl methacrylate
PMP, *see* polymethylpentene
Pneumatic thermoformers, 68, 102
Poly 76, C216
Polyacetal, *see* acetal
Polyacrylonitrile copolymer, T29, 128, T148, 150, T151, T202, C211, C218
Polyallomers, 139, 145
Polyamide, *see* nylon

Polyamide-imide, T29, T202, C219
Polyarylate, C210
Polyarylketone, 145, C220
Polyarylsulfone, 145
Polybutylene terephthalate, 124, 145, C212, C219, C220
Polycarbonate, T8, T16, 21, T29, 37, 49, 71, 114, 120, 127, 141, **142**, 145, T148, 150, 161, 195, T202, T206, C210, C211, C213–215
Polycarbonate sheet, C212, C213, C219
Polycarbonate alloys, C211, C212, C215–220
Polycast Technology Corp., C216, C218, 223
Polychlorotrifluoroethylene, T148, T206, C210, C213
Polydynamics, Inc., 46
Polyester plastic bottle, 123
Polyether-etherketone, T29, 115, T202, C220
Polyetherimide, T29, 143, 144, T202, C219
Polyethersulfone, T29, 143, T202, C217, C220
Polyethylene, 7, T9, T16, 17, 21, 49, 58, 70, 115, 117, **137**, 145, 156, C211, C212, C213, C215, C216
  high-density, T28, T148
  low-density, T28
Polyethylene naphthalate ester, 124, C214
Polyethylene terephthalate (co)polyester, T8, 21, T29, 49, 76, 114, 120, 122–124, **142**, 145, T148, 151, 154, T202, T206, C212, C215–220
Polyfabs, C216
Polyflex, C216
Polyhydroxybutyrate-valerate polymers, 155
Polyimide film or molded parts, C214, C220
Polyketone resin, C211, C214
Polylactic acid based polymers, 155
Polyman C216
Polymer Composites, Inc., C211, 223
Polymer Corp., C213, 223
Polymer Extruded Products, Inc. C214, 223
Polymethyl methacrylate, T8, T28, 37, 76, 114, 121, 122, 129, **136**, 145, 161, T202, T206, C216
Polymethylpentene, T28, **138**, 145, T202, T206, C219
Polyolefins, 3, T28, 37, 76, 117, 119, 127, 134, **137**, T151, T152, 168, T202, C210, C212, C215–218
Polyphenylene ether (oxide), T29, 140, **142**, T202, C215
Polyphenylene sulfide, 115, **143**, 144, T202, C213, C217, C218

238 Index

Polyphthalamide, C210
Polypropylene, 7, T9, 15, 18, 21, T28, 49, 60, 71, 86, 114, 122, **137**, 139, 144, 145, T148, 150, 194, 196, T202, T206, C210, C212, C213, C215–220
Polystyrene, T9, T16, T28, 98, 114, 115, 116, 121, 122, **139**, **140**, 145, T148, 168, T202, T206, C210, C214, C218
Polysulfone, 21, T29, **143**, 145, T202, C215, C217, C219
Polytetrafluorocarbon-coating, 18
Polytetrafluoroethylene, T16, 60, C211, C213, C218
Polytrimethylene terephthalate ester, C211
Polyurethane compounds, C216, C219
Polyvinyl alcohol, T148, 150, T151, T152
Polyvinyl chloride, 4, T8, T29, 49, 67, 76, 114, 115, 117, 127, 134, 138, **140**, 145, T148, 155, 156, 199, T202, T206, C213, C216, C217, C220
Polyvinyl chloride alloy, C212–214, C218
Polyvinyl chloride film and sheet, 128, C214, C217–220
Polyvinyl fluoride, 127, 130, C218
Polyvinylacetate emulsion, C212
Polyvinylidene chloride, 127, **141**, 145, T148, 150, T151, 194, T206, C212, C215, C217
Polyvinylidene fluoride, **141**, 145, T151, C210, C213, C214, C218
Porous metal molds, 56, C215
Portage Industries Corp., 191
Post-consumer generated waste, 154
PP, *see* polypropylene
Preheaters, 138
Preprinted films, 190
Press polishing, 129
Pressure forming, 50, 60, 65, 85, **171**, 198
Prestretching, 37, 59, 64, 164
Prevail, C216
Primacor, C216
Printing, 190
Pro-Fax, C216
Propagation tear strength of films, T206
PS, *see* polystyrene
Pulse, C216
Punch-and-die trimming, 75
PVC, *see* polyvinyl chloride
Pyropel, C217

QLF, 149, C217
Quadrax Corp., C217, 223
Quarite, C217
Quartz, T16
Quartz-like film, 149, C217
QuesTech Packaging Inc., 195
Questra, 122, C217
Quick change locks, 45

Radel, C217
Radiation heating, 3, **5**, 12, 24, 25
Radiation pyrometry, 27
Radio frequency interference, 153
Recycling, regrinding of plastics, 73, 97, 130, 132, **133**, **134**, 152–154
Reflection losses, 7
Regaltech, C217
Regenerated cellulose, *see* cellophane
Repete, C217
Resinite, C217
Retain, C217
Reverse draw forming, 165
Reverse draw with plug-assist forming, 167
Rexene Products Co., C217, 223
Reynolds Metals Co., C217, 223
Reynolon, C217
Ridge forming, 173
Rigidity, 59, 112, 113, 115, 122, **130**, 132, 137, 139, 194
Rigidizing, 187
Rilsan, C217
Riveting, 186
Rohm & Haas Co., 160, 173, 174
Roll-fed thermoformer, 26, 34
Rotary thermoformer, 82, 83
Rotomolding, 114, 199
Rovel, C217
Royalex, C217
Royalite, C217
Royalstat, C217
Rubber diaphragm or pad forming, 175
Rubbery state of plastics, 46, 111, 112
Rynite, C217
Ryton, C217

Sagging of heated sheet, 15, **27**, 30, 83, **112**, 117, 138
SAN, *see* styrene-acrylonitrile copolymer
Sandwich heaters, 3, 15, 18, 79, 116
Santoprene, C217
Saran, T151, 155, C217
Saranex, C217
SB, *see* styrene-butadiene copolymer
Scanning infrared sensor, 25

Schulman, A., Inc., C213, C216, C220, 223
Scrap recycling, 99
Scrapless forming process, 194
Scrapless melt-phase forming process, 195
Segen, Edward, & Co., Inc., 45, C215, 223
Selar, T151, C217
Semicrystalline polymers, 111, 114, 117, **120**, 145
Sewell Plastics Company, 194
Sheffield Plastics, Inc., C213, C214, C220, 223
Shell Chemical Co., C211, C214, C216, C217, C219, 223
Shell Development Company, 194
Shrink film and tubing, 194, C211, C217, C220
Shrinkage due to
   crystallization, 52
   orientation, 121, 128
   thermal contraction, **48**, **49**, 73, **117**
Shuttle thermoformer, 26, 82
Shuttle-mold, 86
Simplicity, 149
Single station machine, 80, 169, 170
Sintra, C218
SIS, T152
Skin packaging, 80, 104
Slip forming, 68, 174
SMA, *see* styrene-maleic anhydride copolymer
Smoke suppressant, 156
Snap fitting, 76, 185
Snap packaging, 104
Soarnol, T152, C218
Softening range, 112
Solacryl, C218
Solef, C218
Solid-phase pressure-forming, 60, 122, 139, 151, **194**
Solmed, C218
Soltan, C218
Solvay Polymers, Inc., C213, C218, 223
Solvent bonding, 186
Sonoco Products Co., 194
Sound barrier sheet, C214
Specific gravity, *see also* density, T16, 122, 137, T202
Specific heat, T16, 17, 21, **115**, 118, 140, T202
Spectar, C218
Spectra, 141
Spectrum, C218
Spring-back, 67, 196
Stacking, 51, 191

Stamping, 68, 144, 195
Stanyl, C218
Stapron, C218
Stat-Rite, C218
Static dissipative plastics, 153, C217, C218
Steel, T16
Steel rule dies, 74
Stefan-Boltzmann constant, 7
Stiffness, *see* rigidity
Stripper plates, 52
Styling considerations, 184
Styrene-acrylonitrile copolymer, T28, **140**, 145, T202, C217, C219
Styrene-butadiene copolymer, **140**, 145, C214, C218
Styrene-isoprene-styrene block copolymer, 152
Styrene-maleic anhydride copolymer, **140**, C211, C212
Styrene-methyl methacrylate copolymer, 140
Styrofoam, C218
Styrolux, C218
Styron, C218
Sullvac, C218
Supec, C218
Sur-Flex, C218
Surface appearance, blisters, detail, texture, 18, 27, 35, **53**, 57, 66, 119, 120, 129, T180, T181, 190, 197
Surface resistivity, *see* electrical properties
Surface temperature, 27
Surge tank, 58, **62**, 64, 109
Surlyn, C218
Syndiotactic plastics, 122, 138, 139, C217
Syntac, C218
Syntactic foam, T16, 60, 165, C218

T-Formcad 5.0, 46
T-SIM 2.2, 47
Tear strength of films, initial and propagation, T206
Tedlar, 130, C218
Teflon, C218
Tefzel, C218
Temperature regulators, 18, 30, 109
Tenite, C219
Terblend, C219
Terluran, C219
Terlux, C219
Testing
   at highest use temperature, 50
   for long-term performance, 131

Tetrafluoroethylene-hexafluoropropylene copolymer, C218
Tetrafluoroethylene-perfluoromethylvinylether copolymer, C213
Texin, C219
Thermal, *see also* heat
Thermal conductivity, 5, T16, 20, 55, **116**, 139, T202
Thermal diffusivity, 71, **118**
Thermal expansion
  coefficient of linear, 16, 18, 48, 49, **116, 117**, 138, T202
  differential, 187
Thermal properties, 15, T16, 130
Thermal stability, 5, 19, 111, **118**, 134, 144
Thermoforming equipment, 70–109
Thermoforming processes, 159–181
Thermoforming window, 27
Thermoset materials, 55, 125
Thermostatically controlled chillers, 70
Thermx, C219
Thickness variations, 3, 12, 30, 39, 41, 48, 51, 59, T180, T181
Thin-gauge
  cooling, 86
  forming, 2, 85
  heating, 22
Thinning of film or sheet, 27, 35–37, 39, T181
Ticona Corp., C213, C219, 223
Tiw-layer resin, T151, C210, C211, C216
Toggle clamp, 65, 68, 103
Tolerances, **48**, 198
Tooling materials, C213, C215, C219, C220
Topas, C219
Toray Plastics (America), Inc., C219, 223
Torayfan, C219
Torlon, C219
Toxicity, 155, 157
Toyo Seikan Kaisha, Ltd., 149, T151, C214, C216, 223
TPX, C219
Transparent materials, 7, 57, 138, 142, 144, 145
Trapped sheet forming, 3, 129, 168
Traytuf, C219
Tredegar Film Products, C215, 223
Triax, C219
Trim press, 91, 93
Trim skeleton, 73
Trimming, 73–78, 80, 83, 85, 87, 103, 168, 190
Troubleshooting, 179, T180, T181

Trycite, C219
Tuffak, C219
Twin-sheet thermoforming, 2, 85, 168, 188, 199
Twinpak Inc., C213, 223
Twintex, C219
Two-dimensional forming, 66, 173
Tyril, C219

Udel, C219
ULDPE, *see* ultralow-density polyethylene
Ultem, C219
Ultimate tensile strength of films, T206
Ultra-high-molecular-weight polyethylene, C213, C214
Ultradur, C219
Ultralite, C219
Ultralow-density polyethylene, 137
Ultramid, C219
Ultrapek, C220
Ultrason, C220
Ultrathene, C220
Ultraviolet radiation protection, 130, 140, 144, C214
Ultraviolet resistance, 144
Undercuts, 51, 52, 185
Unichem, C220
Uniroyal Technology Corp., C212, C217, C218, 223
Unite, T152, C220
Unoflex, C220
Unreacted monomers, 157
Unsaturated polyester resin, 55
Urethane prepolymer adhesive, T151, C214

Vacuum accumulators, 58, **62**, 64, 109
Vacuum distribution system, 54, 62, 63, 109, T181
Vacuum metallizing, 149, 153
Vacuum packaging, 104
Vacuum snap-back forming, 165
Valox, C220
Valtek, C220
Vercon, Inc., 197
Vespel, C220
Vetrotex CertainTeed Corp., C219, 223
Victrex USA Inc., C220, 223
View factor, 12
Vinika, C220
Viscoelastic plateau, 46, 113
Viscous state, 46, 112
VistaFlex, C220

Index **241**

Vitafilm, C220
Vivak, C220
Voids, T180
Volume of air to be evacuated, 58, 62
Vulcan Catalytic Systems, 23
Vydyne, C220
Vynathene, C220

Warm forging, 195
Waste-to-energy conversion, 155
Water, *see also* moisture
Water absorption, **119**, T202
Water jet trimming, 74, 77
Water vapor permeability, 119, 147, 150

Weatherable polymers, 140, 141, C211, C213, C218
Webbing, 27, 41, **43**, 48, 53, 64, 163, T180
Welding, 186
Westlake Plastics Co., C214, 223
Winwrap, C220
Wrinkles, T181

X-TC, C220
Xenoy, C220
XT Polymer, C220
Xydar, C220

Zytel, C220